PRENTICE-HALL, INC., Englewood Cliffs, New Jersey

INTRODUCTION TO ENVIRONMENTAL MICROBIOLOGY

RALPH MITCHELL

Division of Engineering and Applied Physics
Harvard University

Library of Congress Cataloging in Publication Data

Mitchell, Ralph.
    Introduction to environmental microbiology.

    Includes bibliographies.
      1. Microbiology.   2. Microbial ecology.
3. Industrial microbiology.   I. Title.
[DNLM: 1. Air microbiology.   2. Air pollution.
3. Water microbiology.   4. Water pollution.   WA689
M682i   1974]
QR41.2.M57   576   73-19851
ISBN   0-13-482489-X

PRENTICE-HALL SERIES IN ENVIRONMENTAL SCIENCES

Granville H. Sewell, editor

10 9 8 7 6

Printed in the United States of America

PRENTICE-HALL INTERNATIONAL, INC., *London*
PRENTICE-HALL OF AUSTRALIA, PTY. LTD., *Sydney*
PRENTICE-HALL OF CANADA, LTD., *Toronto*
PRENTICE-HALL OF INDIA PRIVATE LIMITED, *New Delhi*
PRENTICE-HALL OF JAPAN, INC., *Tokyo*

For Susan, Tricia, and Dara

CONTENTS

PREFACE

This text is written for students of environmental engineering and management, and the basic sciences. No previous exposure to microbiology is assumed. The book was designed for use in a single semester course in environmental microbiology. The emphasis has been placed deliberately on the human environment, the objective being to make the student aware of the role of microorganisms as pollutants and in pollution control.

I have attempted to synthesize concepts in microbiology, general ecology, and pollution technology, so as to provide the student with a better insight into the activities of microorganisms in a polluted world. The intrusion of technological wastes into the most protected regions of the earth made it imperative that this treatise on environmental microbiology transcend both the pure world of the classical ecologist and the preoccupation with human disease of the classical sanitary engineer. The modern environmental engineer, scientist, or administrator must be capable of understanding basic microbial processes within the context of a contaminated planet. I have, whenever possible, pointed to the use of ecological management practices or technological processes to control environmental deterioration.

The opening chapter provides some of the concepts and language of general ecology. The development of an acceptable human environment is dependent on the application of basic ecological principles. Wherever possible, I have emphasized ecological concepts throughout the book. Roughly, the first third of the book deals with the elements of microbiology. I have included a chapter on waterborne pathogens in this section. The threat of waterborne disease is always present in our developed societies and is still of paramount importance in less developed nations.

Chapters 8 and 9 are devoted to the degradation of both easily utilizable and recalcitrant organic materials, including a detailed critical analysis of the B.O.D. test. These chapters lead naturally into a discussion of nutrient recycling and eutrophication. The transformations of metals were considered to be of sufficient importance to merit a separate chapter. The latter part of the book discusses microbial community ecology and the role of microorganisms in environmental control processes. Biological and advanced waste treatment are treated in separate chapters in order to emphasize the increasing importance of physicochemical processes for pollution control. I have included a chapter on microorganisms as a source of food, since it is apparent that we are moving toward a period of resources recycling, and microorganisms are extraordinarily efficient in the synthesis of protein. The final chapter explains the role of microorganisms in air pollution.

This book should provide the student with all of the information required for a course in environmental microbiology. Obviously, the more adventurous or more advanced person will want to explore further. For this reason I have put a list of references for further reading at the end of each chapter.

I acknowledge with special thanks the many friends and companies who have provided illustrations for this text. I gratefully acknowledge my indebtedness to Professor Werner Stumm of the Swiss Federal Institute of Technology, Zurich. His ideas, expressed during our years as colleagues at Harvard, are reflected throughout the book.

I am grateful to Phoebe Asketh for valuable assistance in the preparation of the manuscript. I also wish to thank Dr. S. Watson of the Woods Hole Oceanographic Institution for the electron micrograph of *Nitrosololobus* on which the dust cover of this book is based.

RALPH MITCHELL

Cambridge, Mass.

ECOLOGICAL
PRINCIPLES

In order to obtain an understanding of the role of microorganisms in the environment, it is essential to know the basic principles of ecology. These provide the framework for a complex network of microbial processes in nature. Microorganisms act as biological catalysts linking the abiotic components of the earth's crust with all other living organisms. Green plants and algae utilize the energy of the sun to transform inorganic materials to building blocks of living cells. These are the prime source of nutrients for all other organisms. Without the action of microorganisms, dead animals and plants would accumulate and choke the surface of the earth. Essential inorganic nutrients for biological growth, such as nitrogen and phosphorus, would rapidly become unavailable.

Perturbations of the soil or water occur by the addition of excessive nutrients, by removal of essential materials, or by the addition of toxic chemicals. Excessive change frequently either stimulates or inhibits one or more microbial processes. A prime example is the runoff of ammonia fertilizer into drinking water reservoirs. Bacterial oxidation of the ammonia yields nitrates at concentrations that are often toxic.

These perturbations are perceived as disturbances in the basic fabric of biological communities. Extreme perturbations may irreversibly destroy the fabric. Maintenance of ecological stability is a prime objective in pollution control.

THE BIOSPHERE The science of *ecology* may be defined as that branch of biology that deals with the interrelationships between living organisms or the interaction between organisms and the abiotic environment.

1

The portion of the earth in which organisms live is called the *biosphere*. The abiotic component of the biosphere includes the inorganic and organic compounds and the climate. The biotic components of the system can be divided into the photosynthetic *primary producers* (the algae and green plants) and the *consumers* (all other living organisms, which ultimately depend on the primary producers for their energy).

Primary Producers   In the open sea, the primary producers are the unicellular algae, the *phytoplankton*, and the multicellular seaweeds. In coastal regions, *benthic* or bottom forms of seaweed are the most important algae. In some parts of the deep ocean, seaweeds may be important producers.

The primary producers in fresh waters are predominantly unicellular algae. In heavily polluted waters, however, aquatic weeds may predominate and shade out the phytoplankton. Green plants are the primary producers in terrestial habitats.

Consumers   The consumers utilize organic sources of food. Their nutrients are the primary producers, excretions of the producers, and organisms or excretions of organisms that feed on the producers. The complex interaction among the abiotic environment, producers, and consumers is shown in Fig. 1.1. When a smaller consumer is eaten by a larger one, each step is a *trophic level*. The producers are at the lowest trophic level. In the oceans, sharks and whales are at the highest trophic level. In the terrestial environment, man is at the highest trophic level.

The Ecosystem   The organisms living together in a specific environment form an *ecosystem*. They depend on each other and on the abiotic environment in which they live. They affect each others' existence in the production of both benevolent and antagonistic materials for other organisms. The biota depletes the biosphere of some materials, and in its excreta and debris it adds other materials.

These tightly meshed interactions define the ecosystem. A lake is an ecosystem in which the abiotic components are temperature and the inorganic compounds in the water column and the sediments. The biota is controlled by the climate and inorganic composition of the lake. Conversely, the biological activity governs the inorganic status of the water.

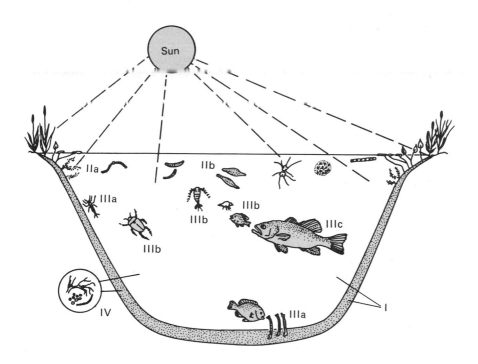

FIGURE 1.1   The complex interaction between the abiotic environment, producers, and consumers: a pond ecosystem. I. Abiotic components. II. Producers: IIa. Rooted vegetation; IIb. Photoplankton. III. Consumers: IIIa. Primary consumers—herbivores; IIIb. Secondary consumers—carnivores; IIIc. Tertiary consumers—secondary carnivores. IV. Microbial decomposers. The system runs on radiant energy. The metabolic rate and stability of the pond depend on the inflow of materials. (After E. Odum, *Fundamentals of Ecology,* 3rd ed. W. B. Saunders, Co., Philadelphia, Pa., 1971.)

THE COMMUNITY

An ecosystem contains a series of populations of organisms living together in a tenuous state of harmony. A *population* denotes a group of organisms belonging to one species. A number of different populations living together is known as a *community*.

The pond in Fig. 1.1 illustrates the delicate balance of an ecosystem that is ultimately dependent on the interaction of many parameters. The abiotic phase contains the inorganic and organic materials necessary for algal growth as well as a light source. The biotic community consists of unicellular algae that grow on the inorganic nutrients using light as their energy source. The consumers are the bacteria and fish.

The sensitivity of the pond community can be shown by observing night-time activity. Primary production stops. Since photosynthesis is accompanied by oxygen release, cessation of algal growth causes reaeration of the water to stop. The fish and bacteria continue to use the dissolved oxygen in the water for their respiration.

An essential population in the community can be suppressed by other forms of perturbation, including chemical pollutants. The destruction of an essential component usually results in deterioration of the community. A restructured community develops that is dominated by a new population, which may not be to our liking. The dominant population in the pond in the absence of algae would be anaerobic bacteria that produce hydrogen sulfide and other noxious gases.

Succession Within communities population changes with time. These changes, which are controlled both by the dynamic nature of the abiotic and biotic components of the ecosystem, are known as *successions*.

In a typical succession the dominant population in the community is continually changing. The new volcanic island of Surtsey, off the coast of Iceland, provides us with an excellent example of succession. This is depicted diagrammatically in Fig. 1.2.

The island first appeared from a volcanic eruption in 1963. Within 6 months, microorganisms (particularly bacteria and fungi) had established themselves in the intertidal zone. A filamentous green alga, *Urosopora*, began to grow at the high tide level in 1965. Within 2 years, 14 species of marine animals had invaded the intertidal zone. In 4 years, 16 algal species had established themselves on the new sea floor. Sea gulls and other birds began to fly over the new island as soon as it was formed. Their feces are providing organic matter and plant seeds for a future plant community.

Climax Communities When the population reaches equilibrium and the communities stop changing, the system has reached a *climax*. We speak of young and old communities. A freshly planted forest is a young community and a meadow is an old fully developed or climax community.

Some ecosystems never reach a climax. The tidal flow in estuaries creates a pulsation and prevents the development of a mature complex ecosystem. Figure 1.3 shows a salt marsh in a

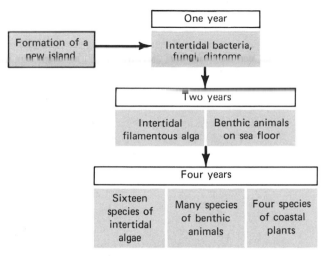

FIGURE 1.2 The ecological succession leading to colonization of the volcanic island of Surtsey, off Iceland. The microorganisms were the first colonists in the intertidal zone. Four years after the formation of the island, not even a microbial community had developed at higher elevations.

FIGURE 1.3 A salt marsh dominated by the marsh grass, *Spartina.* (Courtesy of Massachusetts Audubon Society.)

tidal estuary. The dominant plant is a grass, *Spartina*, which grows in great abundance throughout the marsh. Few other plants are hardy enough to survive the regular tidal flow of the estuary so that a balanced mature plant community never develops. Despite the immaturity of the salt marsh, the *Spartina* population is well entrenched and shows no sign of instability. Indeed, it has been suggested that the salt marsh, because of its ability to tolerate large concentrations of organic matter and low redox potentials, would be a useful place to dispose of domestic sewage.

ENERGY TRANSFER   Organisms living in the biosphere utilize solar radiation and a pool of chemicals in the earth's mantle as their prime sources of energy. This energy is transferred from one organism to another to maintain the complex network of life on the planet.

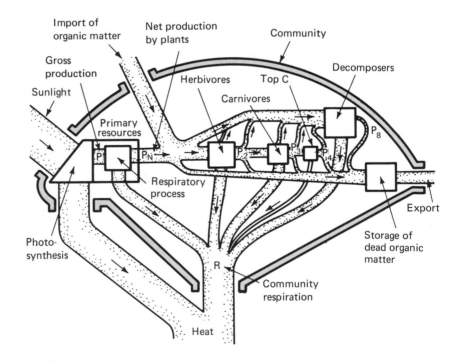

FIGURE 1.4 The food web of a community with a large input of organic matter. There is an extensive loss of respiratory energy (R) at each stage in the process. P is gross primary productivity; P_N is net primary production; P_1, P_2, P_3, P_4, and P_5 represent production at successive trophic levels. (After H. T. Odum, *Limnology and Oceanography*, **1**:101 (1956).)

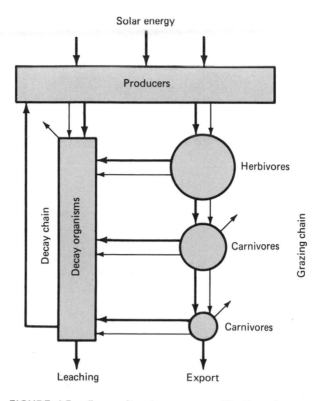

FIGURE 1.5 Energy flow in ecosystems. Net flow of energy (light arrows) and nutrients (black arrows) through a natural community. In a mature community all the energy fixed by the primary producers is dissipated as heat in the respiration of the producers, the consumers (herbivores and successive echelons of carnivores), and decay organisms. Almost all nutrients are eventually recycled, however, to renew producer and consumer populations.

The Food Chain Energy is transferred through the biosphere in *food chains* or, more correctly, *food webs*. The complexity of this web can be seen in Fig. 1.4. The initial fixation of the sun's energy into biological material is carried out by the primary producers. At each trophic level much of the energy is lost as heat. The energy budget is further complicated by microbial degradation. All trophic levels are subject to biodegradation, which may be considered to be a form of grazing. The ecological significance of microbial grazing is unknown.

The flow of energy through a natural community is illustrated graphically in Fig. 1.5. Only a tiny fraction of the solar energy

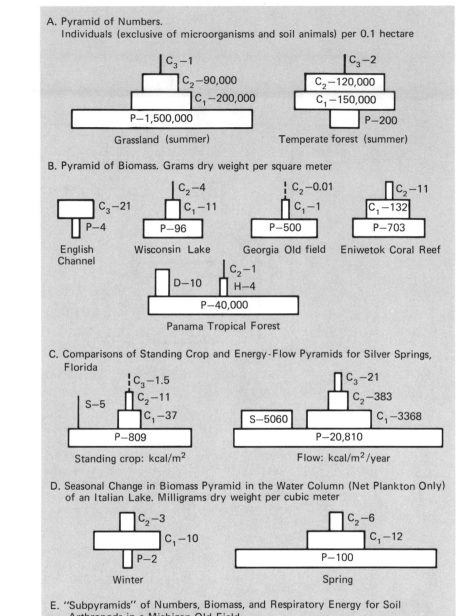

A. Pyramid of Numbers.
Individuals (exclusive of microorganisms and soil animals) per 0.1 hectare

C_3–1
C_2–90,000
C_1–200,000
P–1,500,000
Grassland (summer)

C_3–2
C_2–120,000
C_1–150,000
P–200
Temperate forest (summer)

B. Pyramid of Biomass. Grams dry weight per square meter

C_3–21
P–4
English Channel

C_2–4
C_1–11
P–96
Wisconsin Lake

C_2–0.01
C_1–1
P–500
Georgia Old field

C_2–11
C_1–132
P–703
Eniwetok Coral Reef

D–10
C_2–1
H–4
P–40,000
Panama Tropical Forest

C. Comparisons of Standing Crop and Energy-Flow Pyramids for Silver Springs, Florida

S–5
C_3–1.5
C_2–11
C_1–37
P–809
Standing crop: kcal/m^2

S–5060
C_3–21
C_2–383
C_1–3368
P–20,810
Flow: kcal/m^2/year

D. Seasonal Change in Biomass Pyramid in the Water Column (Net Plankton Only) of an Italian Lake. Milligrams dry weight per cubic meter

C_2–3
C_1–10
P–2
Winter

C_2–6
C_1–12
P–100
Spring

E. "Subpyramids" of Numbers, Biomass, and Respiratory Energy for Soil Arthropods in a Michigan Old Field

C_2–17,000
C_1–142,000
Individuals/m^2

C_2–20
C_1–60
Milligrams/m^2

C_2–625
C_1–1100
gcal/m^2/year

FIGURE 1.6   Ecological pyramids of numbers, biomass, and energy in diverse ecosystems. P = producers; C_1 = primary consumers; C_2 = secondary consumers; C_3 = tertiary (top) consumers; S = saprotrophs (bacteria and fungi); D = decomposers (bacteria, fungi, and detrivores). (From E. Odum, *Fundamentals of Ecology,* 3rd ed. W. B. Saunders, Co., Philadelphia, Pa., 1971.)

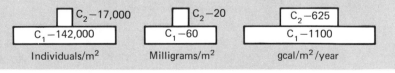

hitting the earth, probably less than 0.1%, is utilized in primary productivity. This figure ultimately is the limiting factor in bio logical productivity in our biosphere. A further loss of between 80 and 90% of the trapped solar energy occurs in transfer to each trophic level. It is easy to see how much more efficient it is to eat rice which is at the first trophic level, than fish, which may be at the fourth or fifth level.

Ecological Pyramids The *ecological pyramid* is an instructive way of looking at energy flow through trophic levels. Eugene Odum defines three types of pyramids:

1. A pyramid of numbers in which the number of individuals are counted at each level.

2. A pyramid of biomass in which the total amount of material, as weight or caloric value, is considered.

3. A pyramid of energy on which the rate of energy flow is depicted.

Odum's three pyramids are explained in Fig. 1.6.

In the first example in Fig. 1.6(a), a large number of organisms from the lower trophic level is required to feed each individual at the next level. Each individual at the third consumer level (C_3) in grassland requires *1½ million* blades of grass. When the individuals consumed are very much smaller than the consumer, this difference may be very large.

The pyramid of biomass is a more useful measure of the flow of energy. It is not necessary to count individuals, and the shape of the pyramid is the same for all systems. Typical biomass pyramids are seen in Fig. 1.6(b). The pyramid is inverted in a highly productive eutrophic water when the standing crop of algae is much greater than that of the animal grazers.

The similarity between biomass pyramids and true energy flow diagrams can be seen from Fig. 1.6(c). When the energy flow is calculated as yield, an almost identical pattern to the one observed using biomass is measured. A serious deficiency in the biomass pyramid shown in Fig. 1.6(b) is the absence of microorganisms. Their importance at the first consumer level is explained in the energy diagram of Fig. 1.6(c). The C_1 level in the biomass diagram represents only snails, insects, and herbivorous fish. The micro-flora comprises a large and very active consumer population degrading the primary producers. The huge loss of energy in the biological use of solar radiation can be seen in the coral atoll example.

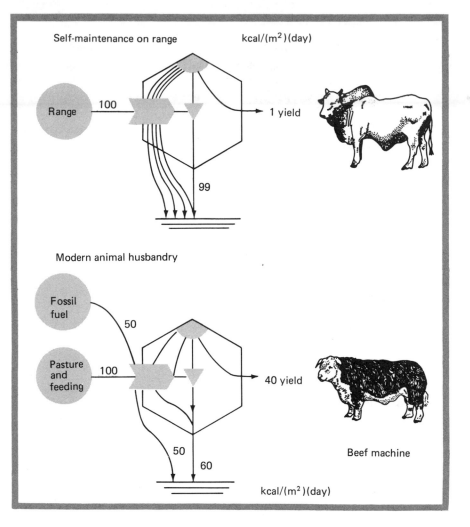

FIGURE 1.7 An example of the effect of fossil fuel energy on agricultural output. Primitive animal husbandry, using the range as an energy source, harnesses a very small percentage of the solar energy. The modern beef machine utilizes the grassland more efficiently and, in addition, utilizes a large external energy input in the form of fertilizers and farm machinery. Fossil fuel is the prime energy source for modern agriculture and industry. (After H. T. Odum, *Environment, Power, and Society.* John Wiley & Sons, Inc., N. Y., 1971.)

The effect of man at the end of the food chain in this natural ecosystem is minimal. The reef supports a little over 100 people with a population density of only 13 per square mile. A very small part of the energy of the reef reaches the human population and their ability to damage the ecosystem is consequently minimized.

Human Inputs   The industrial revolution provided man with a means of subsidizing daily solar energy used in agriculture and fishing. The mining of fossil fuels allows us to utilize solar energy trapped for millions of years to augment our energy supply. The effect of this addition is summarized in Fig. 1.7. Fossil fuels provide the energy for fertilizers and farm machinery. We achieve a higher agricultural yield with the subsidy, although we have not increased the efficiency of energy utilization. This enables us to support a much higher population density. We pay the price in depleted resources and in increased perturbation of the ecosystem because of the release of large quantities of low level energy as heat, animal excreta, or industrial wastes.

Biomagnification   An important attribute of food chains is their ability to concentrate nonmetabolizable toxic materials. This process, known as *biomagnification*, is of great importance in ecology. Materials that are present in extremely low concentration in the abiotic phase can be concentrated in a stepwise manner until at the higher trophic levels they may upset essential metabolic processes. Table 1.1 shows the concentration of DDT residues in a marine ecosystem. The DDT is present in the sea at a concentration in parts per billion. It is concentrated in ever-increasing quantities at each trophic level. The oar weed in the study described in Table 1.1 contained only 0.001 ppm of DDT. Yet the liver of shag in the same area had concentrated 1.56 ppm and the cormorant liver contained 0.19 ppm of DDT.

TABLE 1.1   The concentration of DDT residues in a marine ecosystem.[a]

Species	Trophic Level	DDT (ppm)
Oar weed, *Laminaria*	1	0.001
Sea urchin	2	0.027
Lobster	3	0.024
Shag liver	4	1.56
Cormorant liver	5	0.19

[a]From J. Robinson, et al., *Nature,* 214: 1308 (1967).

The pyramid of numbers is of enormous importance in biomagnification.   The large numbers of individuals consumed to
maintain the next trophic level magnifies the concentration of
toxic material per unit body weight. A pesticide that is present in
the sea at 0.0001 ppm appears in the dolphin four steps later at
1.0 ppm.   Each trophic level magnifies the concentration in the
organisms tenfold (Fig. 1.8).   In addition, the toxin may be
absorbed directly from the aqueous phase.   The ecological danger
of biomagnification is amplified by the concentration of toxic
materials in specific organs and tissues.   The concentration of

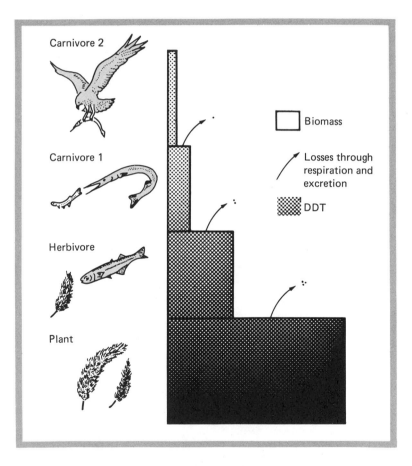

FIGURE 1.8   Biomagnification of DDT. A small amount is lost through respira-
tion at each trophic level. However, the majority is concentrated in new biomass.
(From G. M. Woodwell, "Toxic Substances and Ecological Cycles." Copyright©
March 1967 by Scientific American, Inc. All rights reserved.)

pesticides in fish liver fat tissue and in bird eggs provides a particular danger both to humans and to wildlife.

HOMEOSTASIS   The ability of a system or organism to maintain balance internally against external stress is called *homeostasis*, from the Greek, meaning "same standing." A group of chemicals known as hormones maintain physiological homeostasis in man. In ecosystems this form of self-regulation is achieved by a complex series of abiotic and biotic checks and balances.

Feedback Mechanisms   The basis of homeostasis is in control theory. Control is maintained by achieving a balance between positive and negative feedback. Positive feedback occurs when part of the output is returned to the original input; for example, when part of an algal crop dies, the nutrients return to the water. Negative feedback takes materials out of the system. An example would be the harvesting of an agricultural crop or removal of organic substrates by bacterial growth in a sewage treatment plant. A stylized view of the homeostatic balance between positive and negative feedback can be seen in Fig. 1.9.

FIGURE 1.9 A cybernetic view of the homeostatic balance between positive and negative feedback in a city. (After M. Maruyama, "The Second Cybernetics: Deviation-Amplifying Mutual Causal Process," *American Scientist,* **51,** 1963. Reprinted by permission of the Scientific Research Society of North America, Inc.)

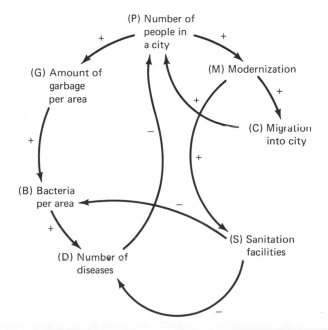

The positive arrow between (G) and (B) indicates that more garbage produces more bacteria. A decrease in garbage would result in a decline in the number of bacteria. The addition of sanitation facilities (S) causes a further decrease in the number of bacteria.

Where mutual causal relationships exist, feedback loops are formed. In the loop P-M-C-P an increase in the population into a city causes modernization that causes further migration into the city and a resultant further population increase. The loop shows that increased population is self-accelerating. Conversely, decreased population in a city leads to deterioration and is also self-accelerating.

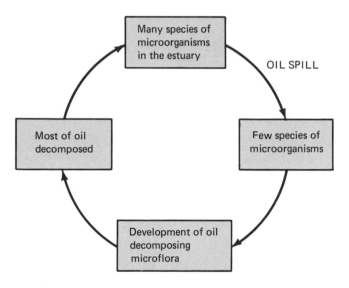

FIGURE 1.10   The homeostatic response to stress in an estuary.

When no external perturbations exist to produce pressure on an ecosystem, the quality and quantity of the biological population is maintained in a stable equilibrium. The presence of external stress tests the homeostatic processes. An example of a homeostatic response to stress is illustrated in Fig. 1.10. The diversity of the flora and fauna of an estuary is seriously depressed when an oil spill occurs. Only those hardy creatures that can tolerate the oil survive and proliferate. Similarly, a large part of the microbial population is eradicated; however, the size of the microbial population does not decline. Hydrocarbon decomposers proliferate and degrade the oil. When the hydrocarbon substrate has disappeared, these microorganisms decline, to be replaced by the diverse normal microbial population. The diversity of the estuarine flora and fauna also slowly increases and returns to its original level.

ECOSYSTEM MANAGEMENT

Man has a tendency to consider the limits of his "oikos", or house, within narrow boundaries, such as his own town or stretch of water. The consequences of this approach in a technological society are disturbing. Our discussion of biomagnification makes it clear that our environment goes far beyond our own small geographical area of interest. The ramifications of environmental perturbation on one side of the world can ultimately be felt on the other side. In order to combat diffuse forms of pollution it is necessary to develop concepts of environmental management.

One approach to ecological control is based on the relationship between the stability of the ecosystem and the broad diversity of

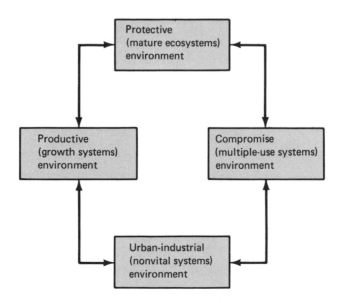

FIGURE 1.11 Ecological management by division of the environment required by man into geographical areas based on agricultural productivity, industry, or recreation. (From E. P. Odum, *Science,* 164: 262 (1969). Copyright by the American Association for the Advancement of Science.)

species of organisms living in that environment. Biological redundancy is an important attribute in maintaining a maximum number of information channels and hence a broad range of response to disturbance.

Eugene Odum has proposed a system of ecosystem management that consciously separates the productive from the protected environment. In Fig. 1.11 there are four areas of management:

1. productive,
2. protective,
3. urban-industrial,
4. compromise.

The size and complexity of these areas would be determined on a regional level to maintain maximum stability.

A major objective of ecosystem management should be the efficient use of an ecosystem to yield maximal productivity. This type of system could involve intensive farming or heavy industry. It is equally important to maintain other ecosystems at low productivity. This includes the majority of our natural waters and the air as potential receiving sources for wastes from the productive ecosystems. Rigid compartmentalization is not possible at present. The ultimate goal, however, must be the recycling of wastes from the productive industrial-agricultural segment of our ecosystem before they reach another nonproductive segment.

SUMMARY

1. Microorganisms are the catalysts linking the abiotic and biotic components of the earth's crust. Perturbations may disturb or even destroy the stability of this complex and finely balanced system.

2. The ecosystem defines the population of organisms living together in a specific environment. The communities change with time until an equilibrium population develops. Pollution prevents the development of a stable ecosystem or disturbs the equilibrium of communities within a stable ecosystem.

3. The energy in the biosphere is transferred to the biota in food chains. The input of energy from fossil fuel results in increased

yields of low level energy to the environment. The energy may be heat, air pollutants, or chemicals released into water.

4. Toxic chemicals that are released at very low concentrations into an ecosystem can be biomagnified in the food chain. This process poses a danger to human health and to wildlife.

5. Cybernetic processes serve to maintain the ecological balance. External stress puts a strain on the homeostatic feedback processes.

6. An understanding of ecological principles provides us with a new approach to pollution control. We can consciously decide whether to utilize an area for high productivity or protect it by applying concepts of ecological management.

FURTHER READING

C. J. Krebs, *Ecology*. Harper & Row, New York, N.Y., 1972.

E. Odum, *Fundamentals of Ecology*, 3rd ed. W. B. Saunders Co., Philadelphia, Pa., 1971.

H. T. Odum, *Environment, Power and Society*. Wiley-Interscience, New York, N.Y., 1971.

"Man and the Ecosphere," *Readings from Scientific American.* W. H. Freeman and Co., San Francisco, Calif., 1971 (paperback).

"The Biosphere," *Readings from Scientific American.* W. H. Freeman and Co., San Francisco, Calif., 1970 (paperback).

"Energy and Power," *Readings from Scientific American.* W. H. Freeman and Co., San Francisco, Calif., 1971 (paperback).

THE PROTISTS

KINGDOMS IN THE BIOLOGICAL WORLD

The living world is divided into three kingdoms: plants, animals, and protists. The plant kingdom can be characterized as the primary producer of the biosphere. Its most important attribute is the ability to utilize the sun's energy to fix atmospheric carbon dioxide into cell tissue. By contrast, animals are consumers. They obtain their energy by utilization of plant material directly or by consuming each other. Both plants and animals are multicellular.

The microorganisms differ from the higher animals and plants because of their simple internal organization. Most are unicellular. The microorganisms have been divided into a separate kingdom, the *protists*. The kingdom includes the protozoa, algae, fungi, slime molds, actinomycetes, blue-green algae, and bacteria. The differences among the three kingdoms are summarized in Fig. 2.1.

The viruses, because of their obligate parasitism, are not classified among the protists. They are living cells, however, and must therefore be considered as microorganisms.

The wide range of size and structure of microorganisms can be seen from Fig. 2.2. Fungi, protozoa, algae, and bacteria are close in size to red blood cells. All are visible in the light microscope. The fine structure of fungi, protozoa, and algae can be studied by light microscopy. Electron microscopy is required to discern the fine structure of bacteria and to see viruses. If we imagine that a virus is the size of an orange, then a large algal cell would be equivalent to a sphere 550 feet in diameter.

At one end of the microbial range are the viruses, which can only divide within the cell of another organism. The algae are at the other end of the range. They require only sunlight and inorganic nutrients for growth. Bacteria and fungi commonly utilize organic compounds in soil and water, in animals, or in plant cells as an energy source. They include *saprophytic* microorganisms which do not cause disease and *pathogens* which live freely but, given an opportunity, will invade humans, animals, or plants and cause disease. The bacterium *Salmonella typhosa* can live outside the human body. When it reaches the human intestine, however, it grows rapidly and causes typhoid. Protozoa may utilize dead organic matter or living bacterial cells or they may be animal pathogens. The vast majority of microorganisms in nature are not pathogenic.

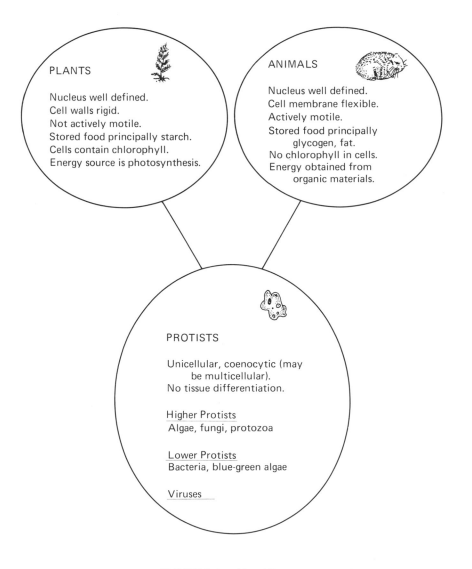

PLANTS

Nucleus well defined.
Cell walls rigid.
Not actively motile.
Stored food principally starch.
Cells contain chlorophyll.
Energy source is photosynthesis.

ANIMALS

Nucleus well defined.
Cell membrane flexible.
Actively motile.
Stored food principally
      glycogen, fat.
No chlorophyll in cells.
Energy obtained from
      organic materials.

PROTISTS

Unicellular, coenocytic (may
      be multicellular).
No tissue differentiation.

Higher Protists
Algae, fungi, protozoa

Lower Protists
Bacteria, blue-green algae

Viruses

FIGURE 2.1   The differences among the animal, plant, and protist kingdoms. The protists are typically unicellular and have a simple internal structure.

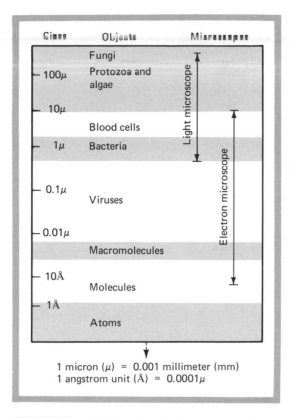

FIGURE 2.2   Relative sizes of microorganisms, blood cells, molecules, and atoms. The visibility in light and electron microscopes is also shown. (After A. J. Rhodes and C. E. van Rooyen, *Textbook of Virology*. The Williams and Wilkins Co., Baltimore, Md., 1958.)

PROCARYOTES AND EUCARYOTES

The protists are divided into two groups based on the degree of differentiation of the cells. The higher group with more differentiated cells is called *eucaryotic*. The lower group with less differentiated cells is called *procaryotic*. The differences between eucaryotic and procaryotic protists are summarized in Table 2.1.

The eucaryotes have a true nucleus similar to animals and plants. The procaryotes do not have a true nucleus. The eucaryotic group includes the protozoa, fungi, and algae. The procaryotic group includes bacteria, actinomycetes, and blue-green algae. The eucaryotic nucleus has a nuclear membrane, undergoes mitosis, and has many chromosomes. The procaryotic nucleus has no membrane, does not undergo mitosis, and has one chromosome.

The cytoplasm of procaryotes is much simpler than the eucaryotic cytoplasm. The eucaryotic cell has specialized structures, such as mitochondria for respiration and an endoplasmic reticulum as an

TABLE 2.1   The major differences between eucaryotic and procaryotic cells.

Characteristic	Eucaryotes	Procaryotes
Nucleus		
Nuclear membrane	Yes	No
Mitosis	Yes	No
Chromosomes	Many	One
Cytoplasm		
Ribosomes	80-S	70-S
Mitochondria	Yes	No
Endoplasmic reticulum	Yes	No
Chloroplasts	In algae	No
Lysosomes	Yes	No
Golgi apparatus	Yes	No
Flagella	Single fibril	Many fibrils
Cell wall	Cellulose and/or chitin	Mucopeptide
Cell size	Greater than 20μ in length	Less than 5μ in length

extension of the cell membrane, lysosomes containing hydrolytic enzymes, and a Golgi apparatus to transport metabolic products (Fig. 2.3). Procaryotic cells have none of these structures (Fig. 2.4).

The flagellum of procaryotes is a single fibril. Eucaryote flagella are composed of bundles of fibrils. The eucaryotic cell wall is composed of either cellulose or chitin. Procaryotic walls are mucopolysaccharides. Eucaryotic cells are typically larger than 20 μ (microns). Procaryotic cells are usually less than 5 μ in size.

Photosynthetic eucaryotic protists have their photosynthetic pigments bound in an organized structure, the chloroplast. Photosynthetic procaryotes carry their photosynthetic pigments in vesicles in the cytoplasm.

The *ribosomes* are particles composed of ribonucleoprotein found in the cytoplasm. They are formed from ribonucleic acid (RNA) and protein. They are essential for the synthesis of protein by the cell. The ribosomes are about 200 Å in diameter and are visible in the electron microscope. Procaryotic ribosomes are distinctly smaller than those from eucaryotic cells. Ribosomes can be differentiated by their rate of sedimentation in a high-speed *ultracentrifuge*. The sedimentation constant S is a function of the size and weight of the particle. Eucaryotic ribosomes have a constant of 80-S, whereas procaryotic ribosomes have a constant of 70-S.

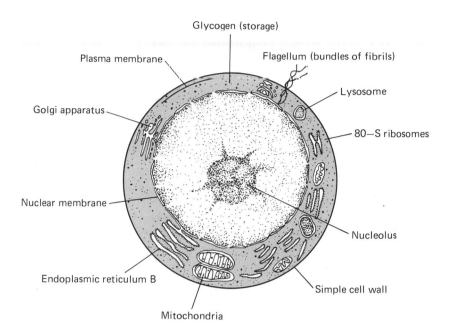

FIGURE 2.3   A diagram of a typical eucaryotic cell.

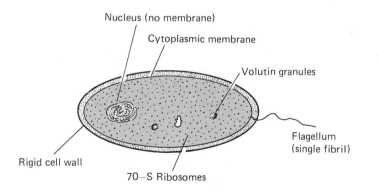

FIGURE 2.4   A diagram of a procaryotic cell.

Eucaryotic cells either synthesize or require the lipid sterol as additives for growth. Procaryotic cells never require or contain sterols.

DISTRIBUTION OF MICROORGANISMS   Most people consider microorganisms as disease-causing agents to be destroyed whenever possible. In fact, the vast majority of microorganisms in natural habitats are not only nonpathogenic but are essential for life in the biosphere. Microorganisms mediate the fixation of light energy in aquatic habitats, degrade organic matter to release essential nutrients, and mediate the oxidation and reduction processes of inorganic compounds.

Huge numbers of microorganisms are present in natural habitats. A gram of soil contains more than 1 million bacteria, and about 10,000 pieces of fungal mycelium. A liter of clean river water contains more than 1 million bacteria, about 10,000 algal cells, and 1000 units of algal filaments. It is apparent that microorganisms are by no means rare in nature and that bacteria are most predominant. We know little about the numbers of protozoa or viruses in soil and water. The common notation for 1 million bacteria per milliter ($\frac{1}{1000}$ liter) is 10^6/ml.

Oxygen Requirement   Most of the protists require the presence of free oxygen in their habitat. These organisms are *aerobes* and they live in an *aerobic* environment. Anaerobic protists require the absence of free oxygen and live in an anaerobic environment. *Clostridium botulinum*, which produces the deadly botulinus toxin and causes the disease botulism, is a strict anaerobe. Humans contract botulism from the consumption of canned goods contaminated with *Cl. botulinum*, from which all O_2 has been expelled. Strict anaerobes predominate in both freshwater and marine sediments and mediate the biochemical processes occurring in these habitats. Many protists are facultative anaerobes. These organisms grow optimally under aerobic conditions. Conversely, one finds facultative aerobes that grow best under anaerobic conditions.

Salt   Many microorganisms living in the sea are unable to grow in fresh water. Marine microorganisms require high concentrations of sodium. This requirement is fulfilled by the 3.5% sodium chloride in the sea. Most marine microorganisms can be adapted to nonmarine habitats and many nonmarine protists can be adapted to a marine environment. The number of obligately marine microorganisms is quite small. Marine protists are moderately *halophilic*, or salt-loving. Extreme halophiles can be found in highly saline environments, such as the Dead Sea or the Great Salt Lake in Utah. These microorganisms are obligately halophilic and cannot be adapted to nonhalophilic conditions. *Halobacterium* is the best known extreme halophile. It is common in the Dead Sea. The effect of salt on growth of halophilic and nonhalophilic protists is seen in Fig. 2.5.

Temperature Most microorganisms living in the soil or in natural waters are *mesophilic*. They grow best at the ambient air temperature of 25°C. Mesophiles will grow to a temperature as low as 15°C and as high as 35°C. Human pathogens are a specialized group of mesophiles adapted to grow at their optimum rate at 37°C, the temperature of the human body. The intestinal bacterium *E. coli* grows optimally at 37°C. Some growth occurs at temperatures as low as 15°C and as high as 44°C.

Psychrophilic, or cold-loving, microorganisms grow best at temperatures just above freezing. Most protists survive freezing but few are adapted to grow at very cold temperatures. There are few true psychrophiles. Protists found growing in very cold environments usually are mesophiles adapted to grow more slowly in the cold. Psychrophilic algae are common and it is quite usual to see a bloom of algae occurring in very cold Arctic water.

Thermophilic, or heat-loving, protists live in compost piles and

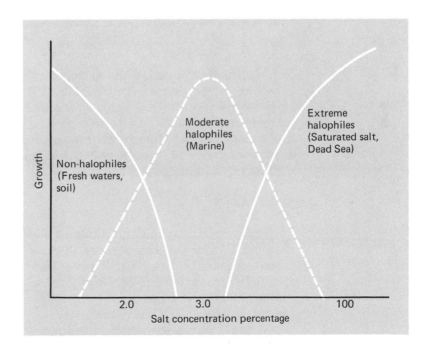

FIGURE 2.5   The effect of salt concentration on the growth of halophilic and nonhalophilic protists.

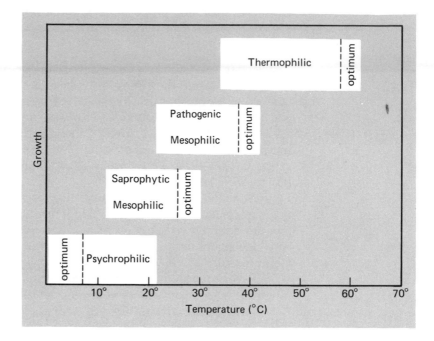

FIGURE 2.6   The growth patterns of psychrophilic, saprophytic and pathogenic mesophiles, and thermophilic protists in relation to temperature.

in hot springs. They have the extraordinary ability of growing optimally at temperatures between 50 and 70°C. Thermophiles usually survive above 100°C. Fungi predominate in compost piles at temperatures of 50–60°C. Bacteria are the predominant protists in hot springs. These bacteria display optimal growth at 70°C. The growth patterns of psychrophilic, mesophilic, and thermophilic protists are summarized in Fig. 2.6.

CULTURE TECHNIQUES

There are many thousand different microorganisms living in the biosphere. In order to study the diseases caused by them or the processes that they mediate, it is necessary to *isolate* and *identify* specific microorganisms. The organisms are grown on sterile *nutrient media* in the laboratory. Sterilization is carried out by heating the nutrients to 120°C for 20 minutes.

Nutrient media are usually maintained free of microorganisms by being covered prior to sterilization with either a cotton plug or a screw cap. Contamination of pure microbial cultures is very com-

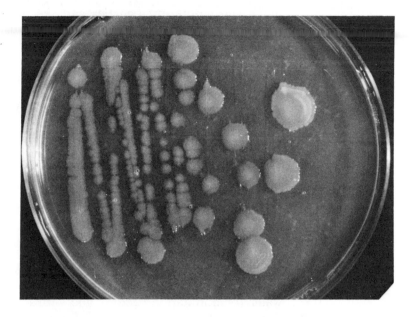

FIGURE 2.7   Bacterial colonies growing following inoculation of bacteria to a solid medium.

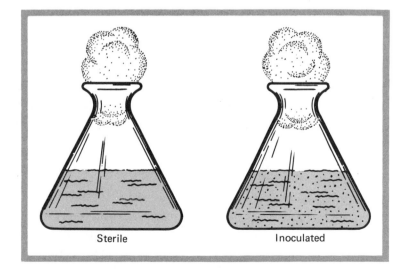

FIGURE 2.8   Bacteria growing in a liquid nutrient medium. The flask on the left contains sterilized nutrients. The culture medium on the right has been inoculated with bacteria and incubated for 24 hours. The turbid flask contains millions of bacteria.

mon. The laboratory benches and the air have a large microbial population. All work with pure cultures must be carried out over a flame and with sterile pipettes. An *inoculation* loop that can easily be flame sterilized is commonly used to transfer or *subculture* microorganisms. This technique yields the bacterial colonies illustrated in Fig. 2.7.

Initial isolation of an organism is usually made in liquid culture. Fig. 2.8 shows a microorganism growing in a liquid nutrient medium. Pure cultures are obtained by isolation of single cells from these liquid nutrient cultures onto sterile media solidified with agar in *petri dishes*. These cells grow into individual colonies (Fig. 2.9).

The nutrient media used to grow microorganisms depend on the nutritional requirements of the organisms. Nutrient broth is a typical commercial medium for the growth of bacteria and is available in dehydrated form. It contains a mixture of proteins, sugars, and growth factors and is designed to fit the nutritional needs of the widest possible group of protists. It can be rehydrated with tap water for nonmarine bacteria and with seawater for marine forms. Commercial media and specific formulations are available for the growth of actinomycetes, fungi, algae, and protozoa.

FIGURE 2.9   A pure culture of bacteria growing on nutrient medium in a petri dish. (Courtesy of Millipore Filter Corp.)

TABLE 2.2   Use of selective media to grow one group of bacteria to the exclusion of others.

Medium	Environment	Selected Bacteria
Lactose Bile salts Broth	37°C Aerobic	Intestinal bacteria (*Escherichia coli*)
Organic media deficient in nitrogen	25–30°C Aerobic	Nitrogen-fixing bacteria (*Azotobacter*)
Organic media containing nitrate	25–30°C Anaerobic	Denitrifying bacteria (*Pseudomonas denitrificans*)
Organic media containing reduced nitrogen compounds and ethanol as the sole carbon source	25–30°C Anaerobic	Methane-producing bacteria (*Methanobacterium*)
Organic media containing reduced nitrogen compounds and glucose as the sole carbon source	25–30°C Anaerobic	Typical anaerobic bacteria (*Clostridium*)

Enrichment Culture   *Selective* media favor the growth of a single organism or group to the exclusion of all others. Table 2.2 shows some of these media.

A combination of lactose and bile salts serves to select or *enrich* for human intestinal bacteria in a water sample. The bacteria utilize lactose as a carbon source and the bile selects for intestinal organisms. Incubation at 37°C further selects for organisms whose normal habitat is the human body.

We can select for the nitrogen-fixing *Azotobacter*, a bacterium capable of growing on organic carbon sources and utilizing atmospheric nitrogen. This organism is common in soil so soil is inoculated into organic media deficient in nitrogen. This medium, incubated in the air at 25-30°C, *selects* for *Azotobacter* and similar organisms.

Organic media containing nitrate, inoculated with soil or water and incubated anaerobically, enrich for *denitrifying* bacteria. These bacteria convert nitrate to nitrogen gas in the absence of oxygen. The methane-producing bacteria are obtained from mixed cultures of bacteria by anaerobic incubation of a medium containing ethanol as a carbon source and reduced nitrogen sources. The anaerobic heterotroph *Clostridium* is obtained following anaerobic incubation when glucose is the carbon source.

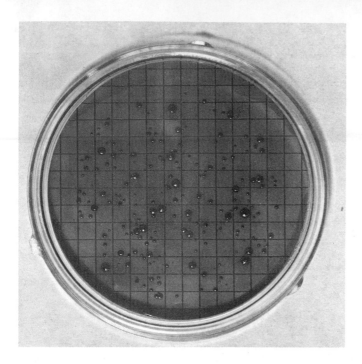

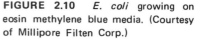

FIGURE 2.10   *E. coli* growing on eosin methylene blue media. (Courtesy of Millipore Filten Corp.)

The enrichment technique is of great importance for the isolation of specific microorganisms from natural habitats. The technique assumes that the organism we are attempting to isolate is present at low concentration in the sample we are studying. One of the most surprising phenomena of microbial ecology is the large diversity of microorganisms in any habitat. Aerobes, anaerobes, and the broadest spectrum of diversity are normally found in any sample of soil or water. The numbers of any one species may be very small. Enrichment culture is used to isolate a single organism or narrow group of organisms from this mixture.

Selective media are particularly useful in medical diagnosis to tell if a specific pathogen is present. A combination of eosine and methylene blue added to lactose agar prevents the growth of most bacteria from food or drinking water. *E. coli* is not inhibited and grows to yield green colonies (Fig. 2.10). This technique is used to prove the occurrence of fecal contamination of food or water. The presence of green colonies on the medium indicates fecal contamination of the sample being tested.

Pure Cultures   When a *pure culture* of the microorganism has been obtained, it is free of all other organisms. This is determined both by microscopic examination and by repeated *subculture* onto fresh sterile growth medium. The same morphological structure and colony shape should be observed in subculture if the culture is pure.

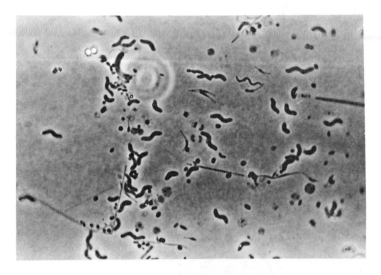

FIGURE 2.11 A mixed population of protists isolated from pond water. Phase contrast microscopy. Magnification 1200X. (Courtesy of P. Hirsch.)

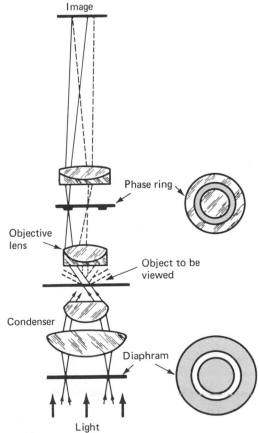

FIGURE 2.12 The principle of phase contrast microscopy. The solid lines show the *transmission* of light through the speci-men and phase ring. The ring diffracts part of the light. The dotted lines show the path of the *diffracted* light. The image that the viewer sees is a combination of transmitted and diffracted light. This combination yields much greater contrast between the phases in the sample. Thus phase contrast microscopy is ex-collent for use with living cells.

FIGURE 2.13   The operation of an electron microscope. A beam of electrons is focused magnetically to yield an image on a fluorescent screen. A camera is mounted below the screen. (Courtesy Jeolco.)

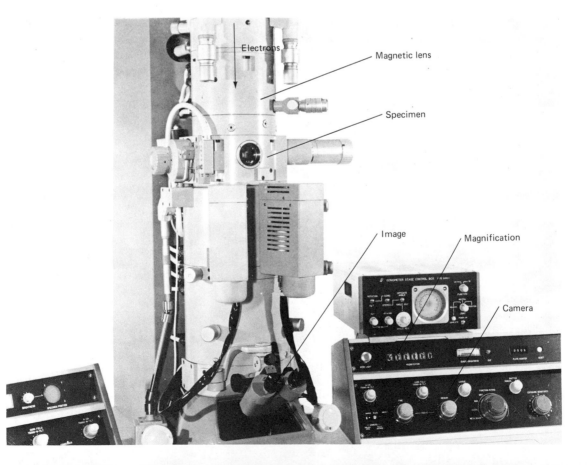

Electrons

Magnetic lens

Specimen

Image

Magnification

Camera

Microscopic Examination   Microscopic observation of microorganisms requires the use of the highest magnification light microscope, yielding a magnification of 1500 times the size of the object. The light microscope only shows objects larger than 0.2 μ in diameter. Viruses are beyond its resolution and cannot be seen. Bacteria whose size ranges from 0.5 to 10 μ, are just within the resolution of the light microscope.

Following heat fixation and staining, protists appear on a microscope slide as distinct shapes (Fig. 2.11). Internal structure is difficult to discern in the light microscope even with the use of differential staining techniques. The living cells of protists are so transparent that they cannot be seen by bright field light microscopy. Figure 2.12 illustrates the principle of phase contrast microscopy used to view living organisms.

Electron Microscopy   The magnification of a cell can be increased to 400,000 using electron beams instead of light. This is the principle of the transmission electron microscope. A photograph of an electron microscope is shown in Fig. 2.13. The cells appear so large that internal structures are apparent (Fig. 2.14).

The scanning electron microscope is a recent innovation. It enables observance of cell surfaces. It is particularly useful for direct observation of samples from natural habitats (Fig. 2.15).

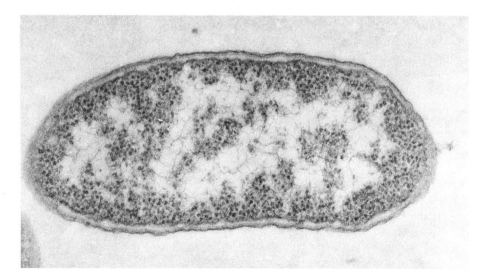

FIGURE 2.14   A transmission electron micrograph of a bacterium. (Courtesy U. Goodenough).

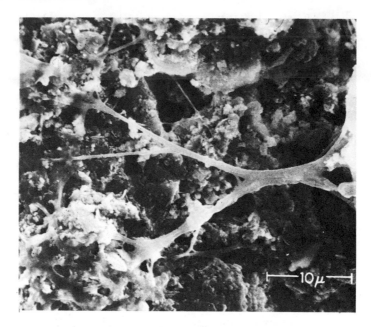

FIGURE 2.15   A scanning electron micrograph of bacteria growing on the surface of coal.

ENUMERATION OF MICROORGANISMS

The number of protists present in a habitat provides an indication of biological activity. The huge populations present in such environments as polluted waters make enumeration difficult. Bacteria account for 99% of the protists present in most environments and much effort is spent in enumerating as many of them as possible.

Microscopic Counts   The simplest method of counting protists is to place a sample on a microscope with squares of a specific area etched on the slide. Slides similar to hemocytometers used for counting red blood cells are used. Figure 2.16 shows algae being counted by this method. It is particularly useful for enumeration of the larger and less common eucaryotes. The procaryotes are frequently too small and the population density too high for accurate analysis.

Dilution Techniques   The dilution technique shown in Fig. 2.17 is most commonly used to count bacteria and fungi. The sample is diluted to extinction in sterile buffer. A typical experiment would

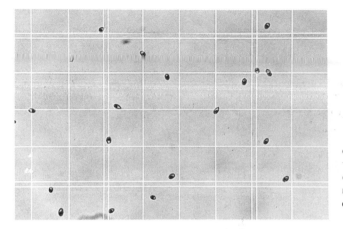

FIGURE 2.16 Counting algae on a hemo-cytometer-type slide. The square bounded by the double lines is 1/25 mm and is 0.1 mm deep. The volume of the square is $1/25 \times 10^{-4}$ ml. There are nine cells in the square. The concentration of cells is $9 \times 25 \times 10^4$/ml or 2×10^6/ml. (Courtesy W. Bell.)

FIGURE 2.17 The dilution count technique for the enumeration of bacteria. Test tubes containing 9 ml of water are sterilized; 1 ml of sample is transferred to the first tube (A) using a sterile pipette. After mixing, the *serial* dilution is continued along the line of tubes; 1 ml is taken from tube A to each of five petri dishes. The procedure is repeated for the other tubes. Melted medium is poured into the dishes. They are mixed and allowed to harden. The dishes are incubated until colonies appear. The colonies are counted. The dishes from tube B average 95 colonies. Those from tube C average 6 colonies. The dilution in tube B is too high to yield an accurate count so dishes from B are discarded. The number of bacteria in the sample is estimated from dilution C as 95×100 or 9.5×10^3/ml of sample.

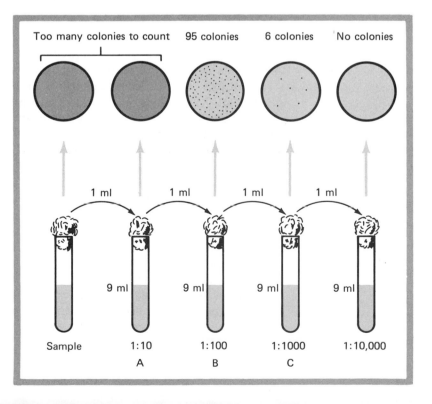

FIGURE 2.18 The Coulter counter utilizes electronic detection to estimate the number of microorganisms in a sample. (Courtesy of Coulter Electronics.)

have 1 ml of a water sample transferred to 9 ml of buffer and mixed. A 1-ml sample from this tube would again be transferred to 9 ml of sterile buffer. Five or six dilutions are made for each sample. Subsamples of 0.1 ml from each tube are spread over plates of a rich nutrient agar medium and incubated to yield colonies. We assume that each colony represents growth from a single cell. Five replicate plates are inoculated from each dilution to overcome variation in the samples. We might find an average of 85 colonies on the five plates each inoculated with 0.1 ml from the 10^{-3} dilution tube. We would estimate that the number of bacteria in the water sample is $85 \times 10^3 \times 10^{-1} = 8.5 \times 10^3$/ml.

The major advantage of the dilution technique is that it yields estimates of numbers of *viable* cells. The microscopic count cannot differentiate between living and dead cells. One critical disadvantage of plate counting is the small number of microorganisms living in natural habitats that yield colonies on conventional media. Nutrient agar is an excellent medium for most pathogens but lacks essential growth factors for many aquatic and soil organisms. We grow soil microorganisms on a medium containing soil extract. Many protists that are catalyzing essential processes in natural waters will not grow on nonspecific organic media. The organisms that are responsible for nitrification are typical of this group

Electron microscopy has shown us that there are many protists living in natural habitats that we are not yet capable of culturing. These were observed by placing electron micrograph grids directly in natural habitats. It is generally accepted that in a typical estimation using conventional plate-counting techniques of a mixed culture of microorganisms from soil or water we only count between 1 and 10% of the protists present.

The Coulter Counter   The electronic Coulter counter (Fig. 2.18) lends itself particularly well to estimations of unicellular algae and protozoa because they are large and have a uniform spherical shape. The counter electronically detects the number of cells within a size range passing across an aperture. It can be used to rapidly determine the total number of algae and protozoa in a sample. If the algae differ significantly in size from the protozoa or if there are two different size groups of organisms, they are counted as separate groups by the counter.

SUMMARY

1. The protists are differentiated by their simple internal organization. They include the protozoa, algae, fungi, slime molds, actinomycetes, blue-green algae, and bacteria. The viruses are not protists.

2. The protists are divided into two groups, the eucaryotes and procaryotes. The eucaryotes have a true nucleus. The procaryotes lack a true nucleus.

3. The eucaryotes include the protozoa, algae, fungi, and slime molds. The procaryotes include the actinomycetes, blue-green algae, and bacteria.

4. Most microorganisms are nonpathogenic. They are found in habitats from boiling hot springs to the Arctic wilderness. Some grow in oxygen, while others are anaerobic. They may use organic or inorganic substrates or they may be photosynthetic.

5. Huge numbers of protists are present in natural environments. They can be separated and enumerated by means of specialized culture techniques.

FURTHER READING

T. D. Brock, *Biology of Microorganisms.* Prentice-Hall, Inc., Englewood Cliffs, N.J., 1970.

M. J. Pelczar and R. D. Reid, *Microbiology*, 3rd ed. McGraw-Hill Book Co., New York, N.Y., 1972.

J. Postgate, *Microbes and Man.* Penguin Books, Inc., Baltimore, Md., 1969 (paperback).

W. R. Sistrom, *Microbial Life*, 2nd ed. Holt, Rinehart and Winston, Inc., New York, N.Y., 1969 (paperback).

R. Y. Stanier, M. Douderoff, and E. A. Adelberg, *The Microbial World*, 3rd ed. Prentice-Hall, Inc., Englewood Cliffs, N.J., 1970.

PROCARYOTES

AND

VIRUSES

The procaryotic group of protists includes the bacteria, actinomycetes, and blue-green algae. They are much simpler micro-organisms than the eucaryotes. As we shall see in this chapter, however, they have a wide range of structure, function, and ecological significance. The viruses, which are neither eucaryotic nor procaryotic, are also considered in this chapter.

THE BACTERIA

Morphology The bacteria are characterized by their small size. They may be as small as $0.5\ \mu$ in diameter and are rarely larger than $10\ \mu$.

The structure of bacterial cells is shown in a stylized diagram in Fig. 3.1. The simplest bacterium has no spores, flagella (singular, flagellum), or capsular material.

All bacterial cells are bounded by a cell wall that maintains the structural integrity of the cell. When the cell walls of rod shaped bacteria are enzymatically digested, the cells revert to spherical forms. The wall also protects the cell against internal osmotic pressure. Cells without walls rapidly *lyse* or burst under the pressure of the cytoplasmic contents. The bacterial cell wall is formed from a highly specific polymer, glycosaminopeptide, and is between 200 and 300 Å thick.

The cytoplasmic membrane is a very thin skin. Its thickness is not more than 75 Å. The membrane acts as a semipermeable barrier protecting the contents of the cell against all components of the ecosystem except nutrients. It allows the passage of waste products out of the cell without permitting the cell contents to escape.

The internal content of the cell, the *cytoplasm*, contains the ribosomes composed of RNA and protein, DNA, and the enzymes necessary for the cell's metabolic processes. The cell stores nutrients as highly refractile *metachromatic granules*. These are predominantly poly-β-hydroxybutyric acid. Starved cells use this stored material to maintain their viability. Figure 3.2 is an electron micrograph of the bacterium *Rhizobium* showing the fine structure.

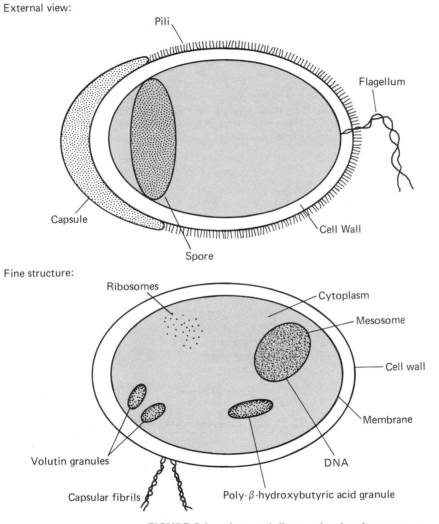

External view:

Pili

Flagellum

Capsule

Cell Wall

Spore

Fine structure:

Ribosomes

Cytoplasm

Mesosome

Cell wall

Membrane

Volutin granules

DNA

Capsular fibrils

Poly-β-hydroxybutyric acid granule

FIGURE 3.1 A general diagram showing the structure of a bacterium. The cell wall provides rigidity to the cell. Motility is usually achieved by flagella, in motile bacteria. The cell membrane inside the wall surrounds the cytoplasm. A capsule may enclose the cell. Nutrients are stored as poly-*β*-hydroxybutyric acid.

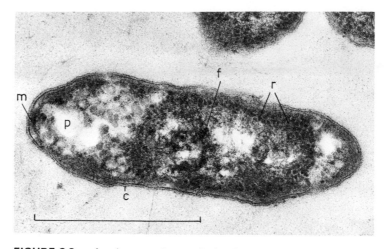

FIGURE 3.2   An electron micrograph showing the internal structure of a typical bacterium, *Rhizobium.* The bar represents 1μ. (c) Cell wall. (m) Protoplasmic membrane. (f) DNA. (r) Ribosomes. (p) Poly-β-hydroxybutyrate granules. (Courtesy A. Bergersen.)

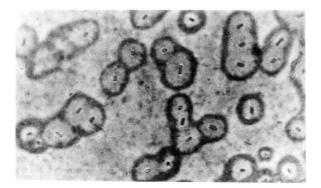

FIGURE 3.3   A bacterium coated with capsular material. The capsule is usually composed of polysaccharide and is extremely mucoid. (Courtesy G. Bitton.)

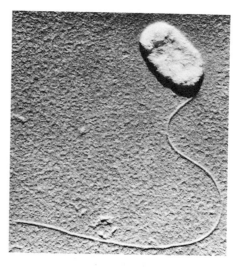

FIGURE 3.4   A bacterium with a single polar flagellum. Magnification 16,000X.

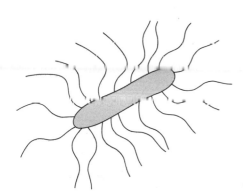

FIGURE 3.5   A bacterium with peritrichous flagella.

A large number of bacteria have short fibers, called *pili*, attached to their walls. The pili apparently are used for attachment to surfaces. They may have an important ecological function in the maintenance of bacteria on surfaces in natural habitats.

Some bacteria produce a thick viscous layer, referred to as a *capsule*, outside the cell wall. Figure 3.3 shows capsulated bacterial cells. Capsules are usually polysaccharides and are continually produced by the cell. The capsule supplements the protection of the cell wall. Pathogenic bacteria often depend on the capsule for infectivity, and loss of the capsule may mean loss of pathogenicity.

Many bacteria are capable of rapid motion in liquid. This *motility* is achieved by means of long thread-like appendages, the flagella. Figure 3.4 is a photograph of a bacterium with a single polar flagellum. Flagella are rarely seen in bacteria other than rod-shaped types. A bacterium may have many flagella attached along its length. These are peritrichous (Fig. 3.5). The presence of flagella and their form are used as characteristics to differentiate genera of bacteria. The flagella are attached to a specialized structure beneath the cyto-plasmic membrane.

Spores Some bacteria are capable of forming resistant bodies, *spores*, that enable the organism to survive extreme heat, irradiation, toxic chemicals, or starvation. When the cell is exposed to adverse conditions, spores form within the cytoplasm. The aerobic bacterium *Bacillus* and the anaerobe *Clostridium* are the most common sporeformers (Fig. 3.6). Both genera are extremely common in soil and water. Their spores can be found in the air and on surfaces. The resistance of the spore is linked to the presence of a cell wall consisting of a dipicolinic acid-calcium complex and to dehydration of the cell contents. When a spore is placed in a nutrient medium and the environmental conditions are not adverse, germination to a vegetative cell occurs. The shape

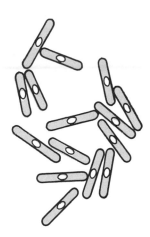
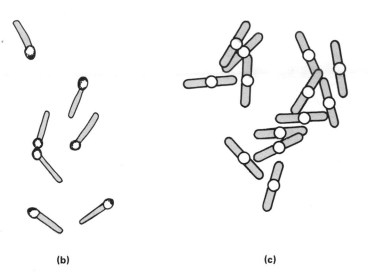

(a)   (b)   (c)

FIGURE 3.6   Spore formation in bacteria. (a) *Bacillus cereus.* Spores are subterminal. (b) *Clostridium tetani,* the tetanus bacterium. Spores are terminal and larger than the vegetative cell. (c) *Bacillus circulans.* Spores are subterminal and larger than the vegetative cell.

and position of a spore within the cell is used for classification. *Bacillus cereus* has a small spore in the center of the cell. *Clostridium tetani* has a large terminal spore.

CLASSIFICATION OF BACTERIA

Morphological Separation   The simplest form of classification of bacteria is based on morphology. Bacteria can be differentiated into three distinct morphological groups (Fig. 3.7): rods, spheres or *cocci* (singular, coccus), and rod-shaped spirals or *spirilli* (singular, spirillum). The genus *Bacillus* is a nonmotile spore-forming rod that occurs in long chains (Fig. 3.8). A species of *Bacillus* is *anthracis. Bacillus anthracis* causes the disease anthrax. Another species, *Bacillus cereus*, is common in soil and does not cause disease. The names of microorganisms are often shortened. We write *B. anthracis* and *B. cereus.* A photomicrograph of *Spirillum,* showing tufts of flagella, can be seen in Fig. 3.9.

We can further differentiate bacteria into either Gram positive or Gram negative on the basis of the Gram stain (Table 3.1). Under the microscope, Gram-positive bacteria appear purple and Gram-negative bacteria appear red. This is a highly reliable test,

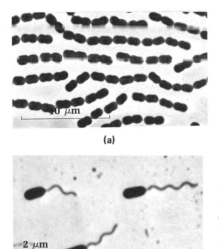

(a)

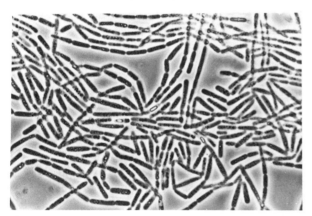

(b)

FIGURE 3.7   Some common bacterial forms. (a) *Streptococcus* forms chains of cocci. (Courtesy of Daisy Kuhn and Patricia Edlemann.) (b) *Pseudomonas,* a short motile rod. (Courtesy of George Hageage and C. F. Robinow.)

FIGURE 3.8   A phase contrast photomicrograph of cells of *Bacillus.* Magnification 1200X. (Courtesy P. Hirsch.)

FIGURE 3.9   Bacteria belonging to the genus *Spirillum,* showing characteristic spiral cells and tufts of flagella. Magnification 1200X. (Courtesy P. Hirsch.)

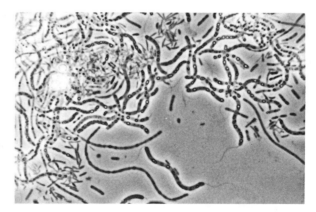

TABLE 3.1   The Gram stain.

Walls of Gram-positive cells form a complex with the crystal violet stain and iodine. This complex makes the cells impermeable to the alcohol rinse, which would wash out the crystal violet color. Hence Gram-positive cells are colored violet. The walls of Gram-negative cells do not hold this complex. Alcohol washes the dye-complex from the walls and they are counter-stained with red safranin dye.

	Bacterial Reaction	
Staining Procedure	*Gram-Positive Cells*	*Gram-Negative Cells*
Crystal violet dye	Cells stain a violet color	Cells stain a violet color
Iodine	A cell-wall iodine-dye complex is formed	
Alcohol rise	Complex, is impermeable to alcohol, violet stain remains	Lipids extracted from cell walls, complex is washed out of porous cell; stain is lost
Safranin	Cells remain violet	Cells counter-stained red

based on cell wall structure. *E. coli* is Gram negative, whereas *Bacillus* is Gram positive. Gram-positive bacteria are susceptible to penicillin. Gram-negative bacteria are not.

Physiological Classification   The separation of the bacteria into taxonomic groups is largely based on their physiological differences. The genus *Cellulomonas*, for example, has as its outstanding characteristic the ability to degrade cellulose. Usually combinations of morphological and physiological characteristics serve to define the taxonomic position of a bacterium. Table 3.2 summarizes some of the best known bacterial genera found in natural habitats together with some of their characteristics.

Bacteria belonging to the genus *Pseudomonas* are among the most common in nature. They are found in soil and water and some species are pathogenic. They are usually pigmented, and none form spores. The placement of bacteria into the genus *Pseudomonas* is based solely on a combination of morphological and physiological characteristics.

By contrast, the genus *Rhizobium* is only found in soil. This bacterium can live symbiotically with leguminous plants to form

TABLE 3.2   Classification of some common bacteria found in natural habitats

Pseudomonas. Short, Gram-negative, nonspore-forming motile rods; aerobic; may produce fluorescent pigments. Most species oxidize glucose to produce acids.

Rhizobium. Gram-negative, aerobic, motile, nonspore-forming rods that form nodules on legumes. Reduce nitrates to nitrites.

Achromobacter. Gram-negative, motile, nonspore-forming rods. No pigments formed.

Flavobacterium. Short, Gram-negative motile rods; produce yellow, red, or orange pigments. Decompose proteins.

Micrococcus. Spherical cells, Gram positive or sometimes Gram negative; cells in irregular groups. All species produce the enzyme catalase.

Sarcina. Spherical cells in regular packets; usually Gram positive; white, yellow, orange, or red pigmentation.

Bacillus. Aerobic or facultatively anaerobic motile spore-forming rods. Decompose proteins to yield ammonia.

Clostridium. Anaerobic spore-forming rods.

nodules and fix atmospheric nitrogen gas (p. 179). It is this ability that is used to separate the genus. *Chromobacterium* is similar to *Rhizobium*, but it is incapable of fixing nitrogen gas and produces a violet pigment.

Numerical Taxonomy The technique of Adansonian analysis is used to classify bacteria using the method of *numerical taxonomy*. A large number of morphological and physiological characteristics are analyzed. Instead of defining the bacterium as a genus or species, it is ranked in degree of similarity to other bacteria using a similarity coefficient:

$$S = \frac{NS}{NS + ND}$$

where NS = number of similar characteristics in two bacteria; ND = number of dissimilar characteristics in two bacteria.

A computer program is utilized to compare the S values among a large number of bacteria. Those bacteria that are very similar to each other are placed next to each other. As similarity decreases, the bacteria are placed farther apart. *Bacillus cereus* and *Bacillus mycoides* in Fig. 3.10 are very similar. This is shown by the dark shading in the area between them. At the other end of the scale *Bacillus cereus* and *Bacillus badius* are very dissimilar. This is indicated by the light shading in the comparison between the two species. The 26 species of *Bacillus* show varying degrees of similarity with each other and these are displayed diagrammatically in Fig. 3.10.

The advantage of numerical taxonomy is that it offers a

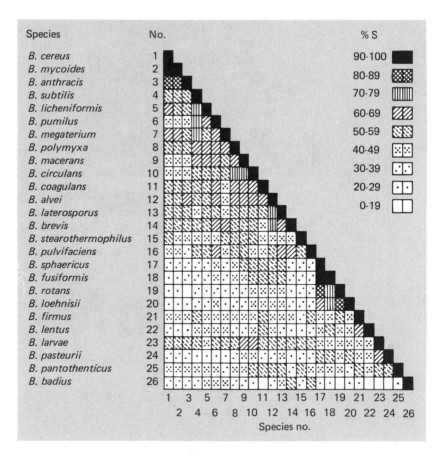

FIGURE 3.10 Numerical taxonomy of 26 species of Bacillus. The matrix shows the degree of similarity among all 26 species. Dark shading indicates great similarity. Light shading represents dissimilarity. (From G. C. Ainsworth and P. H. A. Sneath, *Microbial Classification,* Symposium for the Society for General Microbiology No. 12. Cambridge University Press, Cambridge, Eng., 1962.)

TABLE 3.3   DNA base composition of some common bacteria.

Bacterium	Moles Percent Guanine Plus Cytosine
Clostridium tetani	30–32
Bacillus cereus	34–36
Diplococcus pneumoniae	38–40
Vibrio comma	46–48
Escherichia coli Shigella dysenteriae Salmonella typhosa	50–52
Aerobacter aerogenes	54–56
Pseudomonas fluorescens	60–62
Pseudomonas aeroginosa	66–68

broader, more flexible approach to bacterial classification. There is no constraint to fit a bacterium into a rigid genus and species where it may not belong. Numerical taxonomy allows the conclusion that an isolate of *E. coli* is 80% similar to the *phenotype* or typical form of *E. coli*. The loss of 20% of the characteristics does not force us to look for a new genus or species.

DNA Base Composition   A more technically sophisticated technique of taxonomic analysis is based on the content of the DNA bases guanine + cytosine (GC). The GC content of a bacterial species is highly characteristic (Table 3.3). *Clostridium tetani* is an anaerobic, spore-forming rod that is responsible for the disease tetanus. It contains 30–32 moles % GC. Its closest neighbor is *Bacillus cereus*, which is an aerobic, nonpathogenic, sporeforming rod with 34–36 moles % GC.   The pneumonia bacterium *Diplococcus pneumoniae* contains 38–40 moles % GC and the cholera bacterium *Vibrio comma*, 46–48 moles %.

It is interesting that the three most important enteric bacteria, *Salmonella*, *Shigella*, and *E. coli*, all contain 50–52 moles % GC so they obviously are genetically quite close. *Aerobacter aerogenes* is frequently confused with enteric bacteria in conventional physiological tests. It has quite a different GC ratio and therefore is genetically quite different. The DNA base composition can often be used to differentiate species. *Pseudomonas fluorescens* has 60–62 moles % GC, whereas *Ps. aeroginosa* contains 66–68 moles %.

THE ACTINOMYCETES

Morphology The actinomycetes produce fine filamentous cells, approximately 0.7μ in diameter. Each filament or *hypha* branches to form a mat or *mycelium*. The actinomycetes are classified as bacteria; however, their distinctive morphology and branching growth places them closer to the fungi. They are best considered as a separate group.

Actinomycetes are extremely common in nature. In many natural habitats and particularly in soil and fresh waters the actinomycetes may represent 1% of the protist population or 10,000 units of filaments per gram of soil or milliliter of water. The majority of saprophytic actinomycetes are strict aerobes. Most are mesophilic. *Thermoactinomyces* is a thermophilic form found in compost piles.

CLASSIFICATION OF ACTINOMYCETES

Classification is based on morphology (Table 3.4). Most actinomycetes are sporeformers. The position and size of the spore is used for identification. Two families of actinomycetes are com-

TABLE 3.4 Classification of some common actinomycetes found in natural habitats.

A. Family Actinomycetaceae

Form a mycelium; diameter of hyphae similar to bacteria; hyphae break into fragments yielding bacillary segments and spores

1. *Actinomyces*. Anaerobic; pathogenic, causes human diseases

2. *Nocardia*. Aerobic; very common in water; frequently in oil slicks; good hydrocarbon decomposers

B. Family Streptomycetaceae

Form a true mycelium that does not segment

1. *Streptomyces*. Conidial spores formed at hyphal tips in chains; aerobic and very common in soil and water; many species produce antibiotics

2. *Micromonospora*. A single spore at hyphal tip; found in lake muds

3. *Thermoactinomyces*. A thermophilic genus similar to *Micromonospora*

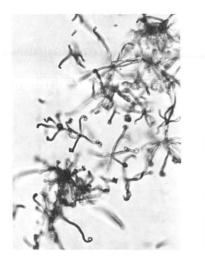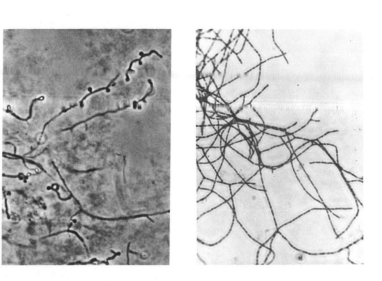

FIGURE 3.11 Structural characteristics of some common genera of actinomycetes. *Streptomyces* is the most common genus found in soil and water. *Micromonospora* is found in lake muds. *Nocardia* is common in habitats containing oil.

mon in natural habitats. The Actinomycetaceae form fine hyphae that segment easily. *Actinomyces* is a human pathogen and is not common in soil or water. Nocardias are very common, however, particularly where hydrocarbons are present.

The Streptomycetaceae are the most common actinomycetes in nature. They are ubiquitous in soil and fresh water. Many of the best known antibiotics, including streptomycin and the tetracyclines, are produced by species of *Streptomyces*. Some common actinomycetes can be seen in Fig. 3.11. Actinomycetes grow much more slowly than the bacteria. It is common to incubate actinomycete cultures for 14 days.

HABITATS

Streptomycetes do not tolerate acidity. They are never found growing in soil or water below pH 5. They will remain viable, however, in the form of spores. *Micromonospora* is found in lake muds. Many species of *Nocardia* develop rapidly in the vicinity of oil spills.

BLUE-GREEN ALGAE

This group of procaryotic protists belongs to the family Cyanophyceae. They differ from the bacteria in their photosynthetic processes. They are obligate photoautotrophs. They utilize CO_2 as their carbon source and sunlight for energy; they use H_2O as an electron acceptor and carry out the classical Hill reaction for the formation of organic carbon compounds:

$$CO_2 + H_2O \xrightarrow{\text{sunlight}} \text{glucose} + O_2$$

This is the same photosynthetic reaction utilized by all green plants and eucaryotic algae. The blue-green algae contain a unique blue pigment, phycocyanin, in addition to the other photosynthetic pigments, chlorophyll a, phycoerythrin, carotene, and xanthophyll.

In all other respects the blue-green algae are structurally similar to true bacteria. They do not have a discrete nucleus. The photosynthetic pigments are not held in chloroplasts. The cell wall structure is bacterial. A comparison between the blue-green algae and eucaryotic algae is shown in Table 3.5.

TABLE 3.5 A comparison of the structure of blue-green algae and eucaryotic algae.

The blue-greens are dissimilar in structure to the true algae. They are physiologically similar in their aerobic metabolism and the use of water as the electron donor.

Characteristic	Blue-Green Algae	Eucaryotic Algae
Chloroplasts	No	Yes
Pigments	Chlorophyll a and b; specific carotenoids; phycobiliproteins	Chlorophyll a and b
Cell wall	Bacterial	Eucaryotic
Electron donor	H_2O	H_2O
Growth	Aerobic	Aerobic

TAXONOMY OF BLUE-GREEN ALGAE

The blue-green algae are classified by means of their morphological structure. They may be coccoid or filamentous (Fig. 3.12). The filamentous forms are all motile. They utilize a gliding mechanism for locomotion. *Anabaena* and *Nostoc* are common coccoid forms. The cells occur in chains. *Oscillatoria* is a characteristic filamentous genus. It grows in long straight filaments. Some common blue-green algae are shown in Fig. 3.13.

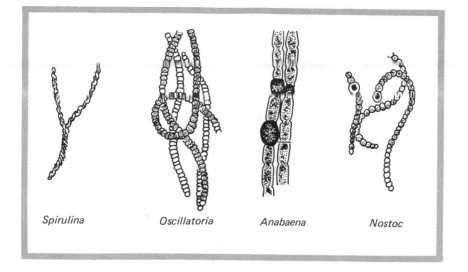

Spirulina   Oscillatoria   Anabaena   Nostoc

FIGURE 3.12   Different morphological forms of blue-green algae.

Blue-green algae are found in both marine and fresh waters. In eutrophic conditions, blooms of blue-greens frequently occur late in the season. This phenomenon is unexplained. However, some blue-greens, particularly the filamentous *Nostoc*, are capable of fixing atmospheric nitrogen. When growth of the eucaryotic

FIGURE 3.13   Some common blue-green algae found in polluted waters. (a) *Anabaena flos-aque.* The dark cell is a resistant heterocyst. (b) *Aphanizomenon* from an algal bloom in a lake. (Courtesy A. Walsby.)

(a)   **(b)**

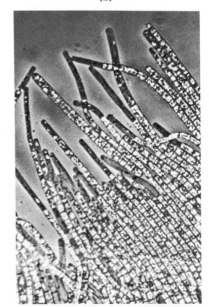

algae is limited by nitrogen deficiency, the nitrogen-fixing pro-caryotic algae predominate.

The Cyanophyceae are pioneers in volcanic and desert areas where no organic matter is present. They grow well in bright sun-light in the presence of an abundance of inorganic nutrients, often at temperatures above 30°C where eucaryotic algae will not grow.

VIRUSES

The smallest microorganisms, the viruses, do not fit into either the eucaryotic or procaryotic group of protists. They differ both in structure and in their reproductive processes from other protists. The viruses are obligate parasites and cannot grow outside a host organism.

Morphology   Poliomyelitis virus is among the smallest viruses. It is 0.25 μ or 25 mμ in diameter. One millimicron equals one millionth of a millimeter. Smallpox virus has a diameter of 250 mμ and is among the largest viruses. Most viruses are beyond the resolution of the light microscope. Cells of smallpox virus appear as barely visible dots. Their extremely small size enables the viruses to be separated easily from other microorganisms. A 0.45-μ pore size filter allows the passage of viruses and holds back all other micro-organisms.

A viral unit, called a *virion*, is formed from nucleic acid and a protein coat or *capsid*. The nucleic acid may be RNA or DNA, and

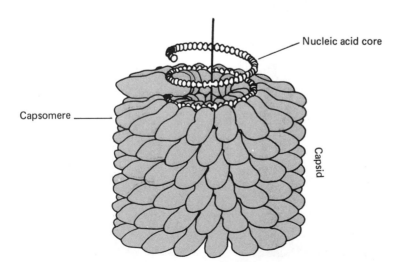

Nucleic acid core

Capsomere

Capsid

FIGURE 3.14   A drawing of a tobacco mosaic virus. Virion show-ing the nucleic acid core and the capsid composed of protein cap-someres. (After A. Klug and D. L. D. Caspar, *Advances in Virus Research,* 7:225. Academic Press, N. Y., 1960.)

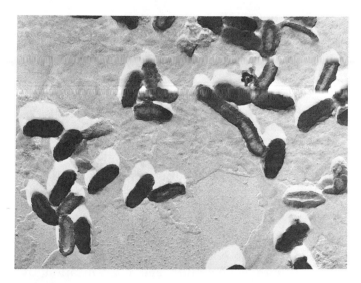

FIGURE 3.15 An electron micrograph of virus. Magnification 57,000X. (Courtesy J. Adams.)

the DNA may be single or double stranded. The capsid is formed from protein units, *capsomeres*. Figure 3.14 provides a diagram of a virion showing the nucleic acid core, the capsid, and capsomeres. Figure 3.15 is an electron micrograph of an insect virus magnified 30,000 times.

Reproduction   Replication only occurs within host cells. The nucleic acid penetrates the cell leaving the capsid outside. Viral nucleic acid appears in the host cytoplasm and provides the template for synthesis of new virions. Viral nucleic acid and capsids are synthesized separately within the host cell. They combine to form mature virions that are released by lysis of the host cell. Bacteriophages are viruses that attack bacteria. They are quite specific. Bacteriophage that attacks *E. coli* will not attack other genera of bacteria. The *coliphages* are used extensively in the study of viruses. Bacteriophages active against many different bacteria, actinomycetes, and blue-green algae have been isolated. Viruses against eucaryotic protists are rare. There have been occasional reports of viral infections of fungi and green algae. Plant viruses multiply within specific tissues of a highly specific host plant.

Animal viruses are quite host specific and cause distinct diseases. Human viruses include smallpox, infectious hepatitis, influenza, yellow fever, and poliomyelitis. In addition, viruses have been implicated in the cause of cancer in humans. The *adenoviruses* are known to cause tumors.

Enumeration   Viruses are enumerated by inoculation to host cells. Bacteriophage is counted by filtering liquid containing the phage to free it of contaminating bacteria. The phage is mixed with host bacteria and placed on a medium on which the host grows well. Following incubation, clearing zones or *plaques* are seen on petri dishes in which the host bacterium has completely covered the dish. Figure 3.16 is a photograph of coliphage plaques on a lawn of *E. coli*. These are easily counted in a manner similar to bacterial colonies. Plant viruses are counted by the number of lesions they form on plant tissues. The technique of *tissue culture* has enabled both enumeration and concentration of animal viruses. Human and animal cells can be cultured in sheets on glass surfaces bathed in nutrient solutions. These tissue cultures are inoculated with human viruses and incubated. Plaques appear when zones of human cells are lysed. Damaged unlysed cells can be detected using stains. Figure 3.17 is a drawing of a tissue culture of animal cells being lysed by a virus.

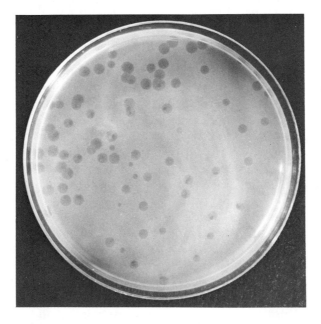

FIGURE 3.16   Clearing zones, or plaques, on a lawn of *Escherichia coli* caused by infection of the bacterium with bacteriophage.

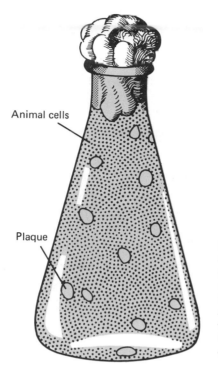

Animal cells

Plaque

FIGURE 3.17 Tissue culture is used to grow and enumerate animal viruses. Animal cells are grown in thin layers attached to the glass surface of a culture bottle. The virus dilution is added to the bottle. Following incubation at 37°C, plaques appear as the cells are lysed by the virus. The virus concentration in a sample can be determined by counting the plaques and multiplying by the dilution.

CLASSIFICATION OF ANIMAL VIRUSES

A classification scheme for some animal viruses is shown in Table 3.6. The primary basis of classification is host specificity and disease epidemiology. Viruses isolated on tissue culture can be identified using morphological and biochemical techniques. Smallpox in humans is caused by the smallpox virus. The cowpox virus causes cowpox in cattle. All the pox viruses are about 250 mμ in diameter. The capsid is coated in an envelope and the nucleic acid is double-stranded DNA. The picornaviruses cause intestinal diseases. One of the picornaviruses, polio virus, causes poliomyelitis in humans. All the picornaviruses have naked capsids. The nucleic acid is a single-stranded RNA.

The paramyxoviruses may be as large as 300 mμ. They are RNA viruses with envelopes and cause mumps and measles in humans. The arboviruses contain an RNA capsid covered with a coat but are very tiny, less than 50 mμ. They cause yellow fever and encephalitis. Tumor viruses have a coated RNA capsid. This type of classification is also used extensively for plant and bacterial viruses.

TABLE 3.6   Classification of animal viruses.

Primary classification is on the basis of host specificity and disease epidemiology. Secondary separation into groups utilizes morphological characteristics.

Virus Group	Size (mμ)	Nucleic Acid	Virion	Typical Diseases
Pox virus	200–250	DNA	With envelope	Human smallpox Cowpox
Picorna-viruses	20–30	RNA	Naked	Poliomyelitis in humans Foot and mouth disease in animals
Paramyxo-viruses	100–300	RNA	With envelope	Measles and mumps in humans
Arbo-viruses	40–50	RNA	With envelope	Arthopod-borne diseases Yellow fever and encephalitis in humans
Tumor viruses	75–125	RNA	With envelope	Murine and avian leukemia Mouse tumors

DISTRIBUTION OF VIRUSES

Despite their inability to multiply outside the host cell and the absence of a resistant body, viruses are capable of survival for long periods in natural habitats. Outbreaks of infectious hepatitis occur from consumption of shellfish contaminated with human feces. Apparently the virus can survive in the sea for extended periods of time. The infectious hepatitis virus cannot be routinely cultured so it is difficult to determine how it survives. However, studies with other viruses have shown that viruses survive well in natural habitats by adsorption to bacterial surfaces and to colloidal debris. Viral contamination of water is discussed in detail on p. 119.

SUMMARY

1. The bacteria are unicellular protists, rarely larger than 10 μ in length. They are common in all habitats. They are bounded by a rigid cell wall and may produce capsules or spores. Some bacteria are motile.

2. Bacteria are classified by morphology and physiological reactions. Modern classification techniques employ either Adansonian or genetic analysis.

3. The actinomycetes are fine filamentous organisms, 0.7 μ in diameter. They are particularly common in soil.

4. Blue-green algae are photosynthetic organisms with a structure identical to bacteria. Their photosynthetic processes are identical to eucaryotic algae and green plants. Many blue-green algae fix atmospheric nitrogen. Both blue-green algae and actinomycetes are classified according to morphology.

5. Viruses are extremely small. Some are no larger than 0.25 μ. They reproduce only within the host cell. Viruses are host specific.

FURTHER READING

T. D. Brock, *Biology of Microorganisms.* Prentice-Hall, Inc., Englewood Cliffs, N.J., 1970. See chapters on procaryotes and viruses.

S. E. Luria and J. E. Darnell, *General Virology*, 2nd ed. John Wiley & Sons, Inc., New York, N.Y., 1967.

M. J. Pelczar and R. D. Reid, *Microbiology*, 3rd ed. McGraw-Hill Book Co., New York, N.Y., 1972. See chapters on procaryotes and viruses.

R. Y. Stanier, M. Douderoff, and E. A. Adelberg, *The Microbial World*, 3rd ed. Prentice-Hall, Inc., Englewood Cliffs, N.J., 1970. See chapters on procaryotes and viruses.

EUCARYOTES

The eucaryotic protists include the fungi, algae, and protozoa. Their structure is much more complex and specialized than the procaryotes. They are also much more diverse. There is little similarity between a green alga and a mushroom.

All eucaryotes have certain characteristics in common (Fig. 4.1). They have a discrete nucleus. Respiratory processes take place in mitochondria. The cell wall is much thicker than in the procaryotes. The walls are relatively simple. They are composed of either a glucose polymer, cellulose, or an N-acetyl glucosamine polymer, chitin. Photosynthetic eucaryotes have their pigments in a distinct body, the chloroplast. Golgi bodies are involved in eucaryotic cell wall synthesis.

FUNGI

Morphology The fungi, or molds as they are sometimes called, are nonphotosynthetic microorganisms that grow by elongation of threads or *hyphae*. These hyphae form a complex branched mass called a *mycelium* (Fig. 4.2). Hyphae are much larger than bacterial cells. The average diameter is 5 μ. Hyphae penetrate into nutrient sources to obtain food. Aerial hyphae usually terminate in spore formations. The green fungus *Penicillium* is well-known for its appearance on stale bread and fruit. It also produces the antibiotic *penicillin*. Figure 4.3 shows a colony of this fungus growing on a nutrient medium. The hyphae have no cross walls. Each hypha contains a mass of cytoplasm and many nuclei. The spores may be either asexual or sexual. The structure of the spore

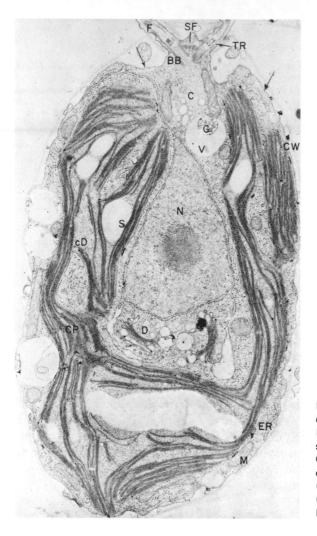

FIGURE 4.1 A photosynthetic eucaryotic cell, *Chlamydomonas,* showing the fine structure. (F) Flagellum; (BB) Basal bodies; (C) Cytoplasmic ground substance (contains ribosomes); (V) Vacuole; (CW) Cell wall; (S) Starch; (N) Nucleus; (CP) Stroma of chloroplast (contains ribosomes); (D) Golgi apparatus; (ER) Endoplasmic reticulum (ribosomes attached); (M) Mitochondria. (From U. W. Goodenough and R. P. Levine, *Sci. Amer.,* **223**; 22 (1970).)

FIGURE 4.2 Mycelium of a fungus, *Trichosporon.* (Courtesy D. Ahearn.)

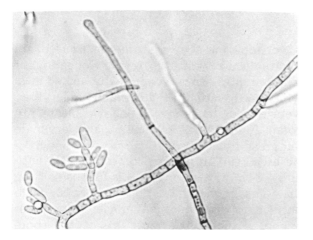

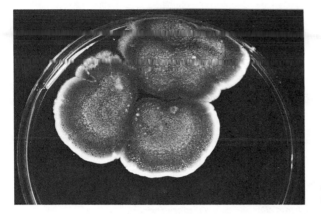

FIGURE 4.3 A colony of the common bread mold, *Penicillium,* well-known for production of the antibiotic penicillin. (Courtesy E. B. G. Jones.)

package and the motility and sexuality of the spores are all used in classification.

Distribution Most fungi are aerobes and are only found in aerobic habitats. They are not active in anaerobic muds or in the deep oxygen-free zones of lakes and oceans. It is difficult to estimate the activity of fungi in soil or water. Each spore and each piece of filament produces a colony on a petri plate. A totally inactive fungal population may produce many thousands of spores. Inoculation of this sample into a nutrient medium induces germination of the spores and a false impression of strong fungal activity. Direct microscopic observation shows whether spores or mycelium predominate in the sample.

Fungi are capable of assimilating an extraordinarily wide range of organic materials. They are particularly adept at utilizing complex carbon compounds and are regularly found growing on cell debris left after bacterial degradation. Synthetic organic chemicals that are disposed into natural environments are frequently utilized as substrates by indigenous fungi. The biodegradation of pesticides and fuel oils by fungi is discussed in Chapter 9.

Cultivation The fungi are easily grown and isolated on nutrient media. They are heterotrophic and can utilize many different organic substrates. They grow best in humid conditions and can survive under strong acidity. Selective media for the growth of fungi frequently have the pH adjusted to 4.0. Almost all fungi are strict aerobes and most are mesophilic with an optimum temperature for growth of 25°C. Pathogenic species have an optimal temperature of 37°C. Thermophilic fungi frequently occur in hot springs and compost piles.

CLASSIFICATION OF FUNGI

The taxonomy of fungi is based on their morphological structure and on their life cycle. Table 4.1 shows the major groups of soil and aquatic fungi and their specific characteristics.

TABLE 4.1   **Typical classification scheme for soil and aquatic fungi.**

There are four major groups divided on the basis of sexual differentiation. Genera are differentiated on the basis of morphology of mycelia and asexual spore formation.

I.   *Phycomycetes.* Sexual spores are free swimming

A. Mucorales. Asexual spore sac is spherical; *Mucor, Rhizopus*

B. Peronosporales. Single, motile, asexual spores; *Pythium*

II.   *Ascomycetes.* Sexual spores in sacs or *asci; Neurospora,* yeasts

III.   *Fungi Imperfecti.* No sexual stage

A. Moniliaceae. Conidiophores in irregular masses; typical genera. *Penicillium. Aspergillus*

B. Dematiaceae. Conidiophores in irregular masses; conidia dark in color; *Alternaria*

IV.   *Basidiomycetes.* Sexual stages borne on special differentiated structures, the basidia; common mushrooms

Phycomycetes   The Phycomycetes are known as water molds. They are relatively simple and have an undifferentiated structure. The life cycle of a water mold, *Allomyces*, is simple and can be seen in Fig. 4.4. Male and female gametes are formed in sacs called *gametangia*. Gametes are motile. *Mucor, Rhizopus*, and *Pythium* are found in soil. *Allomyces* and *Saprolegnia* are aquatic genera.

The phycomycetes can be differentiated by the presence of their asexual spores in a sac, the *sporangium*. Aquatic phycomycetes release motile asexual zoospores from a *zoosporangium*. Terrestial phycomycetes release nonmotile spores. By contrast, asexual spores in the higher fungi are formed singly at hyphal tips. The absence of cross walls in the hyphae serves as another criterion to differentiate the phycomycetes from the higher fungi.

The terrestial phycomycete *Rhizopus* (Fig. 4.5) develops its mycelium into the nutrient source. It moves to new substrates by putting out aerial hyphae, or *stolons*. The spores are borne on sporangiophores

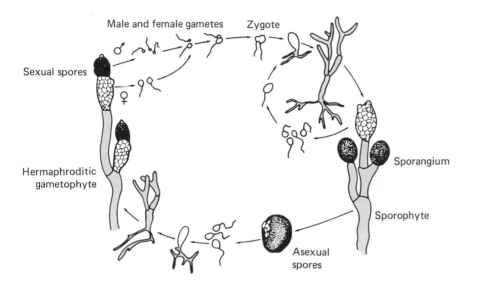

Male and female gametes   Zygote

Sexual spores

Hermaphroditic gametophyte

Sporangium

Sporophyte

Asexual spores

FIGURE 4.4   The life cycle of *Allomyces,* a common aquatic fungus. The fungus reproduces either by asexual spores produced in a sporangium on a sporophyte or by sexual gametes produced in a hermaphroditic gametophyte.

FIGURE 4.5   A diagram of a typical terrestial phycomycete, *Rhizopus stolonifer,* found on rotting fruit and vegetables. The hyphae form a mycelium on the surface and extract nutrients from it. New surfaces are attacked by stolons. The spores are sporangiospores, which are produced in clusters at the tip of a specialized hypha, a sporangiophore.

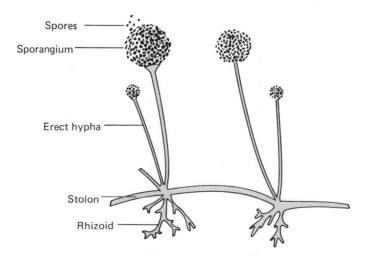

Spores

Sporangium

Erect hypha

Stolon

Rhizoid

Pythium debaryanum is a common plant pathogen. It kills seedlings by producing pectolytic enzymes.

Ascomycetes   The ascomycetes are differentiated from other fungi by carrying their sexual spores in a sac called an *ascus* (Fig. 4.6). The spores are nonmotile. Asexual spores are on hyphal tips but are not always formed. Cross walls, or *septa*, are formed in the hyphae. The septa allow the passage of nuclei and cytoplasmic material.

FIGURE 4.6   An ascus of the marine ascomycete *Phaeonectriella* ascus full of ascospores. (Courtesy E.B.G. Jones and R.A. Eaton.)

Fungi Imperfecti   The *Fungi Imperfecti* or imperfect fungi are ascomycetes that rarely, if ever, produce sexual forms. These are extremely common and account for the majority of fungi found in natural habitats, including *Penicillium*, *Aspergillus*, and *Fusarium*. *Neurospora*, the best known imperfect fungus, can be found but is not well-known in soil or water. Classification is frequently based on the structure and formation of the *conidiophores*, bodies

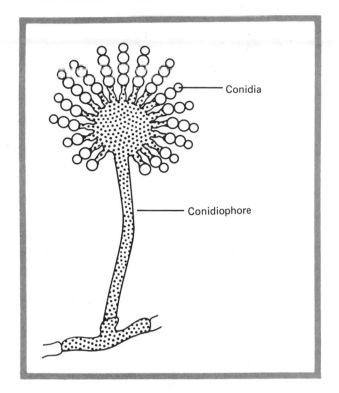

Conidia

Conidiophore

FIGURE 4.7 Fungi Imperfecti reproduce by developing individual spores, conidia, on a fruiting body, the conidiophore.

bearing the asexual spores, the conidia. Figure 4.7 illustrates the morphology of conidiophores of *Aspergillus* and *Penicillium*.

Some of the imperfect fungi are human pathogens. Fungal diseases, or *mycoses*, may be superficial, caused by the fungi growing on the skin, or they may cause deep mycoses in the lungs or other tissues. Athletes foot, caused by *Epidermophyton*, is an example of a skin infection, a *dermatomycosis*. *Candida albicans* is representative of fungi causing deep mycoses. It causes a group of diseases in man and animal of skin membranes and lungs, called *candidiasis*. The fungus is found in feces and is a potential pathogen in contaminated soil. Lung infections can occur by inhalation of dried fecal material containing spores. *Histoplasmosis* is another disease caused by an imperfect fungus, *Histoplasma capsulatium*. This fungus is also carried in feces. Lung infections can occur by inhalation of dust containing spores.

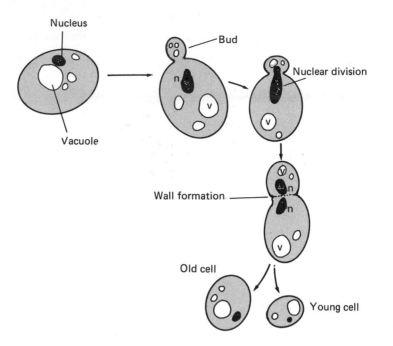

FIGURE 4.8 The development of buds and cell division in yeast cells.

Yeasts The *yeasts* are nonfilamentous fungi. They are unicellular and multiply by budding (Fig. 4.8). Yeasts that produce ascospores belong to the ascomycetes. Nonspore-forming yeasts belong to the Fungi Imperfecti. The yeast used for alcoholic fermentations, *Saccharomyces*, is an ascomycete. *Rhodotorula* is a typical imperfect yeast. *Candida* may occur in either a filamentous

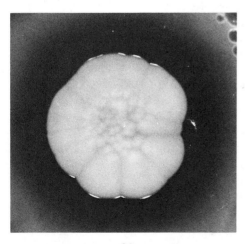

FIGURE 4.9 A colony of yeast growing on agar. (Courtesy D. Ahearn.)

or yeast-like form. A colony of a yeast growing on agar is shown in Fig. 4.9.

Yeasts are very common in soil and water. Often as many as 10,000/ml are found in water. They grow easily on many different substrates. Yeasts tolerate a wide range of temperatures and pH so that it is not surprising to find a high concentration in natural waters.

Basidiomycetes The fourth major group of fungi, the basidiomycetes, or mushrooms, (Fig. 4.10), are highly differentiated. The mycelium is in the soil. In warm, wet weather, specialized structures, the *basidia*, are formed above ground. They eject basidiospores that give rise to more mushrooms so that "colonies" are formed.

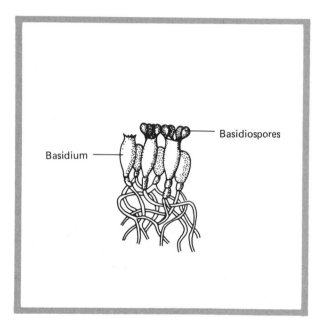

FIGURE 4.10 Basidiomycetes have a specialized spore-forming structure, the basidium. The spores, basidiospores, are formed at the tips of the basidia.

FUNCTION

Fungi are capable of degrading quite complex organic compounds. In soil and water they are found growing on the most resistant fractions of cell debris. The dermatophytic pathogens utilize skin keratin. Plant pathogenic fungi utilize cellulose. Pesticides that are resistant to bacteria are frequently attacked by fungi. Marine forms are common. They grow on cell debris and cause widespread destruction of wood jetties.

PROTOZOA

Morphology   This group of eucaryotes is highly specialized. Many species exist. They are nonphotosynthetic, unicellular organisms that divide by binary fission, and are typically asexual. However, conjugation between cells does occasionally occur. Protozoa do not have true cell walls. They are usually motile, although some species are nonmotile. Under adverse conditions they form cysts with thick walls. Size varies from five microns to hundreds of microns.

Distribution   Protozoa are very widespread in nature. They are found in all soils and waters and in all other habitats where moisture is present. The population of protozoa in soil is frequently between 10,000 and 100,000 per gram. It is not unusual to find 10,000 per milliliter in river water. Sewage is particularly rich in protozoa and may have as many as 1 million per ml.

Cultivation   Most protozoa found in natural environments are predators on bacteria. One finds large protozoan populations wherever bacteria are prevalent. Protozoa are usually isolated, cultured, and counted on media containing large numbers of host bacteria. Most soil or aquatic protozoa grow well on a medium composed of inorganic nutrients enriched with washed cells of a Gram-negative rod. *E. coli* makes an excellent food source. Some protozoa are parasitic on animals. These usually cannot be cultured outside the host animal.

TAXONOMY OF PROTOZOA

The phylum Protozoa can be divided into four subphyla based on locomotion of the cells. The characteristics of these subphyla are summarized in Table 4.2.

TABLE 4.2 **Classification scheme for common aquatic and soil protozoa.**

The protozoa are divided into four subphyla on the basis of their means of locomotion.

I.	*Sarcodina.* Motile by pseudopods; flowing amoeboid motion; *Amoeba, Entamoeba*
II.	*Mastigophora.* Motile by flagella; many are photosynthetic; *Euglena, Volvox, Giardia*
III.	*Ciliaphora.* Motile by many cilia that move in unison; *Paramecium*
IV.	*Sporozoa.* Usually nonmotile; rarely free-living; parasitic

Sarcodines The subphylum Sarcodina comprises organisms that display ameoboid motion. The cell contains a membrane that is not rigid and continually changes shape. *Pseudopodia* or false feet are formed continually in the search for food. A drawing of an amoeba can be seen in Fig. 4.11. Many species of *Amoeba* are found in soil and water. They are all saprophytic. The most common pathogenic amoeba is *Entamoeba histolytica*. This protozoan causes amoebic dysentery in humans. Cysts are found in water contaminated with human feces. However, the amoeba only grows within the human intestine. *E. histolytica* is discussed in more detail on page 120.

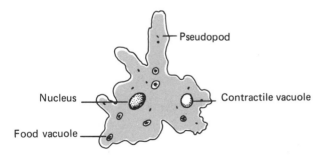

FIGURE 4.11 An amoeba, a member of the subphylum Sarcodina.

Forminifera The forminifera are a subgroup of marine sarcodines that have a shell overlying the membrane. Forminifera are very common in the sea and are not motile.

Mastigophera The subphylum Mastigophera is characterized by the presence of flagella. Division is always longitudinal. Some

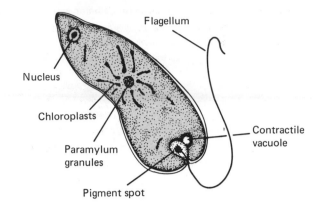

FIGURE 4.12 *Euglena,* a member of the subphylum Mastigophora. This protozoan is photosynthetic and transitional between protozoa and algae.

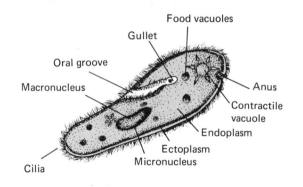

FIGURE 4.13 A member of the subphylum Ciliophora, *Paramecium.* The cilia serve to capture food and for cell motility.

genera are photosynthetic. These organisms are transitional between protozoa and algae. *Euglena* is the most common organism of this group. The diagram in Fig. 4.12 shows that *Euglena* possesses the chief characteristics of both protozoa and algae, being both flagellated and photosynthetic. Nonphotosynthetic mutants of *Euglena* obtain their nutrition by consumption of bacterial cells.

Many Mastigophera are parasitic and do not grow outside the animal host. The best known parasitic flagellate is *Trypanosoma.* This protozoan is the causative organism of sleeping sickness. The tsetse fly is the alternate host.

Ciliates   The ciliaphora, or *ciliates*, achieve their motility by a uniform undulating movement of fine hairs of cilia that surround the cell. The cilia also function in the capture of food. One of the most common ciliates found in water, *Paramecium*, is illustrated in Fig. 4.13. Others include *Stentor* and *Tetrahymena*. Very few ciliates are parasitic.

Sporozoa   The sporozoa are all parasitic protozoa. They are usually not motile. The best known sporizoan is *Plasmodium*, the causative organism of malaria. The alternate host is the mosquito.

FUNCTION

The nonpathogenic protozoa are abundant in natural habitats; yet we know very little about their function. They are voracious eaters of other microorganisms and particularly of bacteria. Protozoa destroy ineffectual native bacteria and nonindigenous bacteria including the huge population in feces entering natural waters or soil. This ability of the protozoa to maintain or restore the biological equilibrium is discussed in more detail in Chapter 13.

ALGAE

Morphology   The algae are a diverse group of photosynthetic eucaryotes. Unicellular algae are called collectively the *phytoplankton*. Multicellular algae, the *seaweeds*, may be many feet in length. *Macrocystis*, the marine kelp found in Pacific coastal waters, is an example. Seaweeds may have holdfasts to grip surfaces. However, they have no specialized structures. They are simply large sheets of the same eucaryotic cells.

The eucaryotic algae have a discrete nucleus. The photosynthetic pigments are held in chloroplasts. All eucaryotic algae contain the pigment chlorophyll a. Green plants use starch as a reserve material. Only the green algae store starch. Others may use glucans or other polysaccharides. Algal cell walls usually contain cellulose and at least two other polysaccharides. Mannans and glucans are common cell wall polymers.

Distribution   Algae require adequate light for growth. They are found in the *euphotic* zone of natural waters. This is the zone where adequate light for photosynthesis is available. In the very clear nutrient-deficient waters of the deep ocean the euphotic zone may be 150 feet deep. Turbid, polluted waters often have a euphotic zone only 1 or 2 feet deep. Algae are also found on the surface of moist soils and rocks. They are present throughout the biosphere. Both thermophilic and psychrophilic algae are known. Many marine algae have temperature optima for growth of 10°C.

The size of the algal population varies tremendously. In deep ocean waters we may find less than 100 algal cells per milliliter. In nutrient-rich coastal waters or in eutrophic lakes the population often reaches 100,000 per milliliter. The algal population in tropical soils may reach 10,000 per gram. In temperate climates the soils have populations of 500 to 1000 algal cells per gram.

Cultivation   Algae excrete large quantities of organic compounds. These excretions, particularly in nutrient-deficient waters, attract a large population of heterotrophic microorganisms. Pure bacteria-free, or *axenic*, cultures of algae are obtained by the judicious use of antibiotics. The correct mixture and concentrations of antibiotics to purify the alga without killing it requires both time and skill. Algae are cultivated on mixtures of inorganic nutrients in the light. Specific formulas having the correct balance of trace elements are required for the growth of each alga. There are few algal media, analogous to nutrient broth, that can be used for the nonspecific growth of mixtures of algae.

The cultivation of algae is further complicated by the inability of some of them to synthesize essential metabolites. These metabolites must be added to the medium as vitamins, or growth factors. In natural habitats these requirements are probably fulfilled by the associated microflora.

CLASSIFICATION OF ALGAE

The algae are divided into groups based on photosynthetic pigments. A simplified classification scheme is summarized in Table 4.3.

Green Algae   The largest division of algae is the Chlorophyta, or *green algae*. This group is characterized by the presence of chlorophyll a and b. The cell wall is cellulose and the storage material

TABLE 4.3   Classification of algae.

Five groups can be differentiated on the basis of pigmentation.

	Group	Pigments	Other Characteristics	Typical Genera
I	Chlorophyta Green algae	Chlorophyll a and b	Unicellular, coenocytic, or multicellular	*Chlorella*, unicellular *Ulva*, multicellular
II	Chrysophyta Diatoms	Chlorophyll a and c Xanthophylls	Unicellular, filamentous, or coenocytic, silaceous cell wall	*Asterionella*
III	Pyrrophyta Dinoflagellates	Chlorophyll a an c Specific carotenoids	Most unicellular, motile by two flagella	*Noctiluca*
IV	Phaeophyta Brown algae	Fucoscanthin	Multicellular; most marine	*Fucus*
V	Rhodophyta Red algae	Phycocyanin	Some unicellular, most multicellular; most marine	*Polysiphonia*

is starch. Green algae may be unicellular or filamentous. Motile cells have two identical flagella. Both marine and freshwater forms are found throughout the world. The green algae *Dunaliella* and *Chlorella* are shown in Fig. 4.14.

Diatoms   The chrysophyta are unicellular golden-brown algae containing chlorophyll a and c and xanthophylls. The diatom wall has quite a characteristic morphology and usually contains silica. The morphology of the wall is frequently used in the classification of diatoms. The reserve material is laminarin. Diatoms live well in cold waters. It is common to find them predominant in Arctic waters and during the winter months in more temperate waters. Some common diatoms are shown in Fig. 4.15.

Dinoflagellates   The Pyrrophyta, or *dinoflagellates*, are unicellular algae containing chlorophyll a and c together with carotenoid pigments (Table 4.3). The cell wall is cellulose and the storage material is starch. They are motile and are separated from other unicellular motile algae by the presence of two flagella, one long and the other short.

The dinoflagellates are as important in warm waters as the diatoms are in cold water. Dinoflagellates are the dominant phytoplankton in tropical marine waters. They are differentiated on the basis of their highly characteristic cell wall morphology. Figure

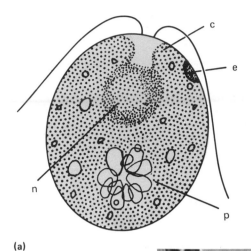

(a)

FIGURE 4.14   (a) The green alga *Dunaliella salina.* This alga is motile by two flagella. (c) Chloroplast; (e) Eye-spot; (p) Starch grains; (n) Nucleus. (b) A photomicrograph of *Chlorella.* (From W. C. McElroy and C. P. Swanson, *Foundations of Biology.* Prentice-Hall, Inc., Englewood Cliffs, N. J., 1968.)

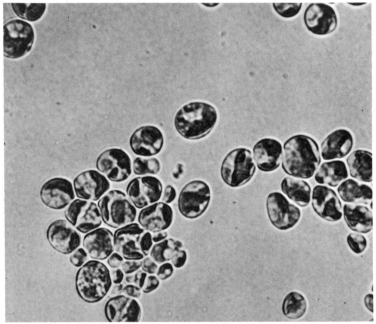

(b)

4.16 illustrates the cell wall structure of *Gonyaulax* and Fig. 4.17 shows the wall structure of *Prorocentrum. Gonyaulax* is responsible for the red tides frequently observed in warm waters. It produces an endotoxin that is responsible for the fish kills caused by the red tide. Another dinoflagellate, *Noctiluca*, develops beautiful luminescent blooms. Dinoflagellates are also associated with corals. The *zooxanthellae* of corals are dinoflagellates of the genus *Zooxanthella*, which live symbiotically on the coral.

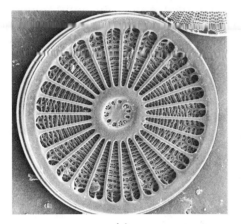

(a)

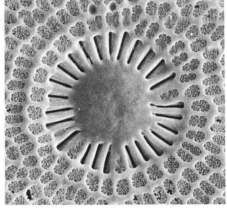

(b)

FIGURE 4.15 Electron micrographs of diatoms. (a) *Arachoidiscus,* magnification 450X. (b) The cell wall structure of *Arachoidiscus,* magnification 3000X. (c) *Navicula,* magnification 3000X. (Courtesy Jeolco Corp.)

(c)

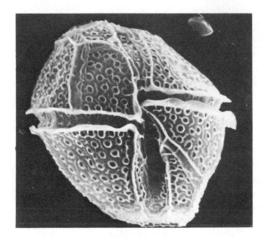

FIGURE 4.16 A scanning electron micrograph of a dinoflagellate, *Gonyaulax,* which causes red tides. Magnification 2000X. (From A. F. Loeblich, III, *Proc. N. Amer. Paleont. Conv.,* 867 (1969).)

Brown Algae   The Phaeophyta, or *brown algae*, are multicellular seaweeds.   They contain chlorophyll a and c and specific carotenoids.   The cell wall contains cellulose and alginic acid.   Storage materials include a glucose polymer laminarin.   The brown algae produce both asexual and sexual spores and have a complex life cycle.   This group of algae is mainly marine and includes the important genera *Fucus* (Fig. 4.18), *Macrocystis*, and *Laminaria*. These algae form large beds in coastal waters and are important breeding and feeding areas for coastal fish.

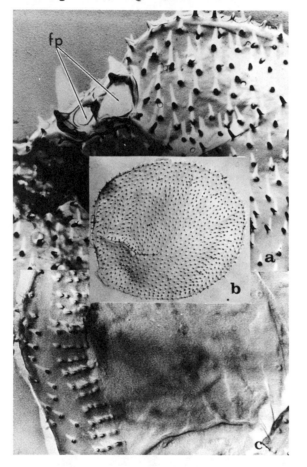

FIGURE 4.17   Electron micrographs of a common dinoflagellate, *Prorocentrum*. (a) The cell wall showing the pores for the two flagella (fp), magnification 13,000X. (b) The external surface of one of the two valves, magnification 3800X. (c) The margin between the two valves, magnification 11,000X. (From A. R. Loeblich, III, *Proc. N. Amer. Paleont. Conv.,* 867 (1969).)

FIGURE 4.18 A brown alga. The seaweed *Fucus*. (Courtesy General Biological Supply House, Inc.)

Red Algae   The Rhodophyta, or *red algae*, contain chlorophyll a and a red pigment, phycobilin. The wall is cellulose and the storage material is starch. Both unicellular and multicellular forms are common. Most red algae are marine. They are very common in subtropical waters where they live in the *subtidal* or totally submerged zone. They do not survive well in tidal areas. *Polysiphonia* is a common red alga.

FUNCTION   The algae are the major primary producers of organic matter in aquatic habitats. Their photosynthetic ability enables them to utilize the energy of the sun to synthesize cellular material. Primary productivity is discussed in more detail in Chapter 11.

SUMMARY

1. Fungi are heterotrophic filamentous microorganisms with a diameter of about 5 μ. Masses of hyphae form a mycelium. They are common in all habitats. Fungi are classified according to morphology and life cycle.

2. Protozoa are heterotrophic unicellular eucaryotes without cell walls. Some are as small as 5 μ in diameter, while others may be 100 μ in diameter. Many protozoa are predators on bacteria.

Others are parasitic on animals. They are classified by morphology and life cycle.

3. Algae are photosynthetic eucaryotes. They may be unicellular or multicellular. Algae are found in all natural waters where light penetrates. They are classified by morphology and life cycle.

FURTHER READING

C. J. Alexopoulos, *Introductory Mycology*, 2nd ed. John Wiley & Sons, Inc. New York, N.Y., 1962.

R. Lewin (ed.), *Physiology and Biochemistry of Algae*. Academic Press, New York, N.Y., 1962.

R. P. Hall, *Protozoology*. Prentice-Hall, Inc., Englewood Cliffs, N.J., 1953.

R. Y. Stanier, M. Douderoff, and E. A. Adelberg, *The Microbial World*, 3rd ed. Prentice-Hall, Inc., Englewood Cliffs, N.J., 1970. See chapters on eucaryotes.

NUTRITION

AND

GROWTH

5

Microorganisms assimilate materials directly from the abiotic portion of an ecosystem or from materials released by excretion or death of other organisms. These materials are utilized as energy sources and as building blocks for microbial growth. In the process a new microbial population is developed and *catabolites*, or degradation products, are excreted. This chapter describes the nutrition and growth of microorganisms and the process by which they obtain energy.

NUTRITION

A symbiotic relationship exists in an ecosystem between the abiotic pool of nutrients and the biota. Microorganisms are dependent on the nutrient pool for growth. At the same time they are constantly recycling nutrients to replenish the abiotic pool.

The microbial cell contains a wide variety of elements in a surprisingly constant ratio. We usually assume that the cell contains approximately 50% carbon, 5—15% nitrogen, 0.5—1.5% phosphorus, and 0.5—1.5% sulfur. The C:N:P:S ratios of the cell are thus 100:10:1:1. In addition, the cell contains hydrogen, oxygen, potassium, calcium, magnesium, sodium, iron, and minute quantities of other elements. The elements that are *limiting* in the ecosystem are most important in our studies of cell growth. By "limiting," we mean that the element is not present in sufficient concentration to allow the organism its full growth range. The three elements in natural ecosystems that frequently limit growth are carbon, nitrogen, and phosphorous because of the high concentration in the cell. Frequently maximal growth cannot be achieved because one of these elements is deficient. The relationship between limiting nutrients and algal productivity is discussed further in Chapter 11.

Growth Factors   A microbial population can be expected to develop in a system containing a range of carbon compounds together with a surplus of inorganic minerals. The complex communities found in soil and natural waters are continually exchanging nutri-ents with the abiotic pool. Many of these microorganisms are incapable of synthesizing organic compounds essential for their growth. These materials, known as *growth factors* or *vitamins*, must be supplied to the cell in a complete form.

The requirement for growth factors is highly specific (Table 5.1). Yeasts require an external source of pyridoxine, which is also known as vitamin B_6, for their amino acid metabolism. *Clostridium butyricum* cannot synthesize biotin, which is necessary for carboxylation reactions. The lacctobacilli require a number of different vitamins for essential metabolic reactions. *L. casei* requires riboflavin (vitamin B_2) for electron transport. Many different bacteria need cobamide (vitamin B_{12}) for molecular rearrangement reactions. *L. arabinosus* must have an external source of panthothenic acid for its fatty acid metabolism.

Precursors   A *precursor* is a chemical that is used for the synthesis of another chemical. Growth factors are usually precursors of

TABLE 5.1   **Some examples of vitamin requirements by yeasts and bacteria and the essential function of the vitamin.**

Bacterium	*Vitamin*	*Function*
Yeasts	Pyridoxine (vitamin B_6)	Amino acid metabolism
Clostridium butyricum	Biotin	Carboxyl metabolism
Streptococcus fecalis	Thiamine (vitamin B_1)	Decarboxylation reactions
Lactabacillus casei	Riboflavin (vitamin B_2)	Electron transport
Lactobacillus	Cobamide (vitamin B_{12})	Molecular rearrangements
Lactobacillus arabinosus	Panthothenic acid	Fatty acid metabolism

essential metabolites and can be replaced by chemicals produced by their reaction. Many *E. coli* mutants lack the capacity to produce essential amino acids, which are precursors of polypeptides and ultimately proteins. The addition of the polypeptide to a medium can substitute for the precursor. Biotin is essential for the synthesis of aspartic acid. Those microorganisms that cannot synthesize their own biotin require it as a growth factor. Aspartic acid can be substituted for biotin.

In natural habitats, microorganisms depend on excretion by neighbors or on cell lysis of associated microorganisms for their growth factors. This aspect of community ecology is discussed in more detail in Chapter 13.

HETEROTROPHY AND AUTOTROPHY

The protists can be divided into three groups on the basis of carbon and energy source:

1. heterotrophs,
2. photoautotrophs,
3. chemoautotrophs.

Heterotrophs   A heterotrophic organism uses organic carbon compounds both as carbon building blocks and as energy sources. Most organisms including all animals and the majority of microorganisms are heterotrophic. *Aerobacter aerogenes* is a typical example of a heterotrophic bacterium. It can use many different carbon compounds as the sole source of carbon. The wide diversity of organic compounds on earth provides a vast pool for the growth of heterotrophs. The range of organic compounds available is matched by the extraordinary range of microorganisms. Transient microbial populations develop in response to the release of each product of catabolism of an organic compound into the environment. These populations are replaced by others as one carbon substrate disappears and another takes its place.

Photoautotrophs   A photoautotroph utilizes carbon dioxide as its sole carbon source and sunlight as its energy source. The green algae are eucaryotic photoautotrophs. Procaryotic photoautotrophs include the blue-green algae and photosynthetic bacteria. The procaryotic photoautotrophs have their pigments associated with the cell membrane rather than in a discrete chloroplast. They also have characteristic pigments.

TABLE 5.2   **Major characteristics of the photosynthetic bacteria.**

They are all anaerobic and use H_2S or organic materials as their electron donors. Two groups are differentiated on the basis of pigments.

  I.   *Green Bacteria.* All photoautotrophic, although they can assimilate acetate; *Chlorobium*

 II.   *Purple Bacteria*

     A. *Thiorhodaceae.* Use H_2S as electron donor

         1. Deposit elemental sulfur outside cells; *Ectothiorhodospira*

         2. Deposit elemental sulfur in cells; *Chromatium, Thiospirillum, Lamprocystis*

     B. *Athiorhodaceae.* Non-sulfur-purple bacteria; use organic electron donors; *Rhodospirillum*

Photosynthetic Bacteria   The photosynthetic reaction of the blue-green algae is identical to that of the green algae (p. 192). In contrast, the photosynthetic bacteria are usually strictly anaerobic and the electron donor is not water.

The photosynthetic bacteria use either hydrogen sulfide or reduced organic compounds as their electron donors (Table 5.2). The green bacteria are obligately photoautotrophic. *Chlorobium* is a characteristic genus.

The photosynthetic reaction of *Chlorobium* is

$$CO_2 + H_2S \rightarrow (CH_2O) + S^0$$

The purple photosynthetic bacteria are divided into two groups. The sulfur-purple Thiorhodaceae use H_2S as their electron donor. *Ectothionhodospira* deposits sulfur outside the cells (Fig. 5.1). *Chromatium, Thiospirillum* (Fig. 5.2), and *Lamprocystis* (Fig. 5.3) deposit sulfur inside the cells.

The second group of purple bacteria, the Athiorhodaceae, are known as *non-sulfur-purple* bacteria. They use organic electron donors. *Rhodospirillum* is the best known of the Athiorhodaceae.

The eucaryotic photoautotrophs, the green algae, are distributed through the world's natural waters and on soil surfaces. The only limitation to their growth is light. The blue-green algae, despite their physiological similarity to their eucaryotic cousins, are not so ubiquitous. They are present in small numbers in natural waters. They are rarely dominant except in very enriched ecosystems and then only at very high light intensity and elevated temperatures. Blooms of blue-green algae typically occur in eutrophic waters at the peak of the summer season.

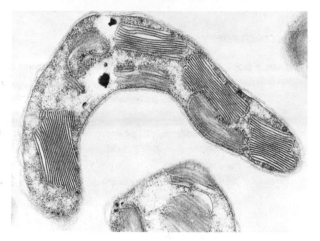

FIGURE 5.1 An electron micrograph of the photosynthetic bacterium *Ectothiorhodospira,* a sulfur-purple bacterium. The photosynthetic lamellae are coiled in sheets. Magnification 40,000X. (From C. C. Remsen, S. W. Watson, J. W. Waterbury, and H. G. Truper, *J. Bacteriol.,* **95**: 2374 (1968).)

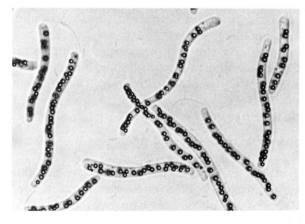

FIGURE 5.2 The sulfur-purple bacterium *Thiospirillum* showing sulfur globules in the cells. Magnification 1200X. (Courtesy N. Pfennig.)

FIGURE 5.3 The photosynthetic sulfur-purple bacterium *Lamprocystis* showing intracellular sulfur globules and gas vacuoles. Magnification 2000X. (Courtesy N. Pfennig.)

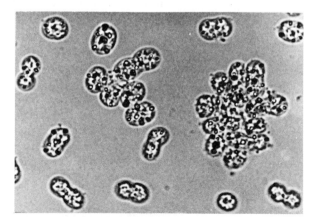

The photosynthetic bacteria are found in places where oxygen is deficient but light is not a limiting factor. Figure 5.4 is a photomicrograph of the spiral sulfur-purple bacterium *Chromatium*. This and other photosynthetic bacteria can be found in the sediments of ponds, in estuarine sediments, and in a narrow bank at a depth of approximately 80 to 150 feet in deep lakes and in the open ocean where oxygen has been depleted and there is still a low light intensity.

Chemoautotrophs   The chemoautotrophic bacteria utilize CO_2 as their sole carbon source. Their energy comes from the oxidation of reduced inorganic compounds. The sulfur bacteria utilize reduced sulfur compounds as energy sources. *Thiobacillus thiooxidans* oxidizes elemental sulfur to sulfate. The nitrifying bacteria oxidize ammonium to nitrate:

$$NH_4^+ \xrightarrow{\text{\textit{Nitrosomonas}}} NO_2^- \xrightarrow{\text{\textit{Nitrobacter}}} NO_3^-$$

These are *obligately* autotrophic bacteria. Many bacteria are *facultative* chemoautotrophs. They can use either an inorganic or an organic energy source. The hydrogen bacterium *Hydrogenomonas* is typical of this group. It can use hydrogen gas as its sole energy source. In the absence of H_2 it can grow well on many organic compounds.

The chemoautotrophs together with the photoautotrophs are the *primary producers* of organic material in the biosphere. They probably were among the first organisms on our earth because of their ability to tap the energy stored in simple inorganic compounds. We find chemoautotrophic bacteria growing in rocky areas devoid

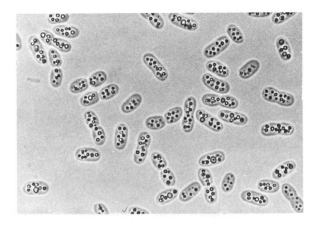

FIGURE 5.4   A photomicrograph of the photosynthetic sulfur-purple bacterium *Chromatium*. Magnification 2000X. (Courtesy N. Pfennig.)

of organic matter. They are also abundant in the open ocean and in unpolluted fresh waters. It is possible to differentiate between uncontaminated and polluted waters by measuring the ratio of chemoautotrophic to heterotrophic bacteria. We return to this subject in more detail in Chapter 14 when we discuss the effect of stress on microbial communities.

ENERGY TRANSFER

The microflora utilizes the available abiotic pool as a source of energy and building blocks for growth. The metabolism of these nutrients by microorganisms can be divided into the following:

1. *Catabolic* processes. Substrates are utilized and decomposed by the cell. In the process, energy is released for later use by the cell. The combined reactions leading to substrate decomposition and energy release are referred to as *catabolic*. They are *exothermic*, or energy producing.

2. *Anabolic* processes. These reactions are utilized by the cell for the synthesis of new cellular material. The combination of synthetic reactions are referred to as *anabolic*. These processes are *endothermic* or energy consuming.

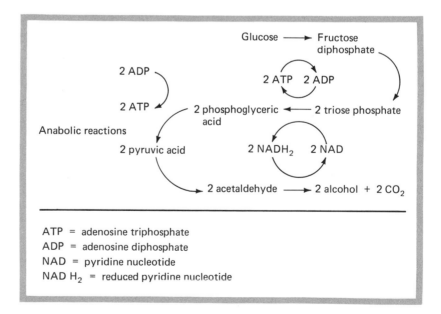

ATP = adenosine triphosphate
ADP = adenosine diphosphate
NAD = pyridine nucleotide
NAD H$_2$ = reduced pyridine nucleotide

FIGURE 5.5 Coupling of catabolism and anabolism in microorganisms. Two molecules of ATP are generated in the catabolism of glucose to alcohol. These are sources of energy for anabolic reactions.

Adenosine Triphosphate (ATP) One of the most important intermediates in the transfer of energy in biological systems is adenosine triphosphate (ATP). The ability of ATP to store energy depends on the formation of two energy-rich phosphate bonds. When energy is required for biosynthetic processes, one or both of the high energy phosphate bonds are broken to yield adenosine diphosphate (ADP) or adenosine monophosphate (AMP). Eight thousand calories (8 kcals) are released for each mole of phosphate released. A typical coupling of catabolic and anabolic processes that is very common in microorganisms is shown in Fig. 5.5. The production of pyruvic acid by the decomposition of glucose yields 2 moles of ATP with 16 kcal of stored energy for use in anabolic processes. Pyruvic acid is a pivotal intermediate in metabolic processes. Under aerobic conditions the most efficient processes would yield CO_2 and 38 ATP molecules. Under strict anaerobic conditions no oxidation occurs and the total yield of ATP is only 2 molecules. In practice the substrate is rarely either completely broken to CO_2 or metabolized under totally anaerobic conditions.

We can divide the use of energy-yielding substrates of the cell into two groups on the basis of the mechanism of energy release:

1. respiration,
2. fermentation.

Respiration This is the most efficient means of utilizing energy. Either organic or inorganic substrates are consumed by aerobic microorganisms. These are oxidized with the release of electrons and the production of ATP from the energy of oxidation. The ultimate electron acceptor is oxygen, which is reduced to H_2O.

$$Glucose + O_2 \rightarrow H_2O + CO_2 + 38 \text{ ATP}$$

Fermentation In fermentative processes, organic compounds are oxidized to release electrons and produce energy (exothermic reactions). The electron acceptors are other more oxidized organic compounds that are reduced with the consumption of energy (endothermic reactions). The balance between exothermic and endothermic reactions results in a net yield of energy; however, the yield is much less than in respiratory processes. The end products are mixed and not necessarily fully oxidized. The pathways depend on the substrate and on the microorganism utilizing it.

Glucose yields many different products when utilized as a substrate by different microorganisms (Fig. 5.6). The key intermediate is pyruvic acid. The lactobacilli convert the pyruvic acid to lactic acid. This is the reaction governing the souring of milk. Yeasts produce ethyl alcohol. This reaction is used to produce alcoholic

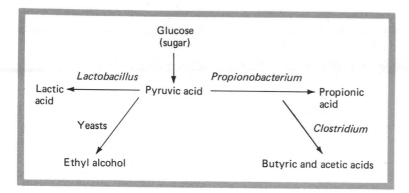

FIGURE 5.6 Examples of common pathways of sugar fermentation by bacteria.

beverages. Propionic acid is formed by *Propionobacterium*. Butyric and acetic acids are produced by the strict anaerobe *Clostridium*. *Anaerobic* fermentations do not use O_2 as the ultimate electron acceptor in contrast to *aerobic* fermentations.

It should be apparent that fermentation is the major process used by microorganisms for the transfer of energy from substrate to cell. The fermentation industry utilizes this information for the production of many products, including alcoholic beverages, dairy products, and many pharmaceutical agents. The ecological significance is great. The nature of the substrates and the oxygen regime of the habitat determine the metabolic rate, the nature of the microflora, and the products produced.

GROWTH KINETICS   Bacterial populations attain high densities very rapidly. Bacteria reproduce by binary fission. Each cell doubles at a rate that is characteristic for that organism. A rapid population explosion is produced (Fig. 5.7). *E. coli* divides approximately every 30 minutes. Inoculation of 1000 cells of *E. coli* to a nutrient medium will yield more than 1 million cells within 6 hours.

The Growth Curve   When an inoculum of bacteria is placed in a new medium, growth follows the curve shown in Fig. 5.8. There are four phases in the bacterial growth curve:

1. lag,
2. exponential,
3. stationary,
4. death.

The first or *lag phase* of growth is a period of adaptation of the cells to a new environment. Frequently the cells in the lag phase are smaller and not yet ready to divide. The exponential phase is not reached until a majority of the cells have reached adult size and have begun to divide.

At this point the population explosion referred to above begins. Each individual cell is doubling every few minutes. While some cells are dying during the exponential phase of growth, the rate of cell multiplication far exceeds the death rate. The growth rate in this phase is constant under constant environmental conditions. The slope of the curve is characteristic for each bacterium.

The exponential phase finishes when either a substrate becomes limiting or a toxic metabolite accumulates in the system. At this point in growth the rate of death equals the growth rate. This is the stationary growth phase. In the final death phase the population declines because cells are dying faster than they are multiplying. The bacterial growth curve must be considered as a statistical phenomenon. Any single cell at any point on the curve may be out of phase with the statistical population.

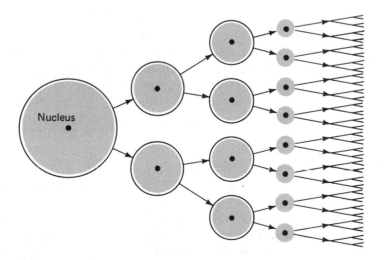

FIGURE 5.7 Multiplication of microorganisms by binary fission. The result is a rapid population explosion. A million bacteria may be produced within hours.

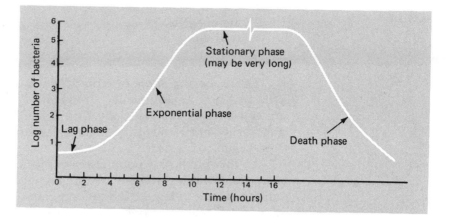

FIGURE 5.8   A typical bacterial growth curve showing the lag, exponential, stationary, and death phases. The number of bacteria is typical. The time may be much longer however, for slow growing bacteria.

Kinetics   The *growth rate* of a bacterium can be calculated if we knew the number of cells in the medium at two times during the exponential growth phase.   At times t_1 and t_2 the population of a bacterium X is a_1 and a_2. The relationship between a_1 and a_2 may be expressed as

$$a_2 = 2^{k(t_2 - t_1)} a_1$$

where k is a constant; the growth rate for bacterium X is expressed as the number of doubling times per hour.

The equation is normally expressed

$$k = \frac{\log_2 a_2 - \log_2 a_1}{t_2 - t_1}$$

The doubling time for a bacterium is $1/k$.

The growth rate constant k assumes no substrate or environmental limitation of growth. k values are usually obtained in a *closed system*, usually a flask of nutrient medium incubated under optimal growth conditions in the laboratory. In natural habitats or in biological treatment facilities, such as a sewage treatment plant, the system is *open* and growth is usually limited to a level below the maximum for that organism.

The Chemostat It is possible to study growth kinetics in *open fully mixed* systems. The *chemostat* (Fig. 5.9) is an apparatus in which cells can be maintained in the exponential phase without reaching the maximal growth rate. The growth rate is expressed as a function of the rate of flow of nutrient medium through the reaction chamber.

This change can be expressed as a function of dilution (*D*) that is equivalent to the reciprocal of the substrate concentration (*S*).

$$\frac{dN}{dt} = \mu N - DN$$
$$= (\mu - D)N$$

where *N* equals the number of bacteria.

This expression says that the growth rate μ of a cell is inversely proportional to the dilution rate (*D*) or is proportional to the substrate concentration (*S*). The growth rate reaches a maximum

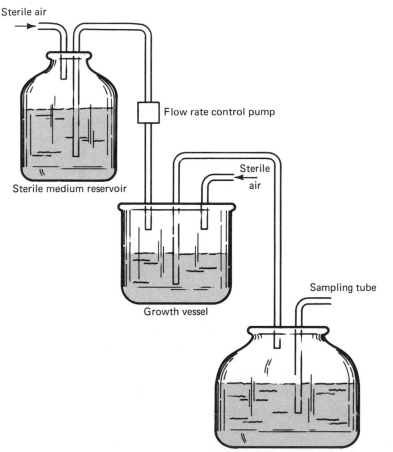

Sterile air

Flow rate control pump

Sterile medium reservoir

Sterile air

Sampling tube

Growth vessel

Washed out cells and spent medium

FIGURE 5.9 A chemostat for the continuous culture of bacteria.

(μ_{max}) at a specific dilution rate or substrate concentration that is characteristic for that organism. The flow rate is adjusted so that the growth rate matches the rate of washout of cells from the growth chamber. There is a value of $1/D$ below which μ reaches zero, indicating that bacterial growth requires a minimal substrate concentration. In the deep ocean where food is very scarce there are places where the substrate concentration is too low for most bacteria to grow.

Unicellular algae divide by binary fission. Their growth kinetics are, therefore, the same as the kinetics of bacterial growth. Similar growth curves are observed and generation times are calculated using the same equations. Algae are commonly grown in the chemostat.

Fungi grow in three dimensions and do not follow bacterial kinetics. The kinetics of fungal and protozoan growth have not been developed. Neither fungi nor protozoa are routinely grown in chemostats.

MEASUREMENT OF GROWTH

The study of bacterial growth kinetics requires us to measure regularly the density of the bacterial population. Four methods are commonly used to assay growth:

1. plate counts,
2. optical density,
3. dissolved oxygen,
4. protein concentration.

FIGURE 5.10 A standard curve used to relate the number of bacteria to the optical density of the suspension.

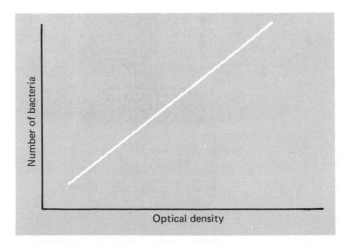

Optical Density   The plate-counting technique described on p. 35 — is time-consuming. Kinetic studies require measurement of growth at frequent intervals. Plate counting would also require large amounts of material. The optical density of a pure culture of bacteria is a direct and reproducible measure of the number of cells. A standard curve can be drawn relating the optical density of a cell suspension to the number of bacteria in a plate-counting assay (Fig. 5.10). In a kinetic study, optical density measurements can be made at regular intervals and read off the standard curve as number of cells per milliliter or used directly. This method is not applicable to mixed cultures.

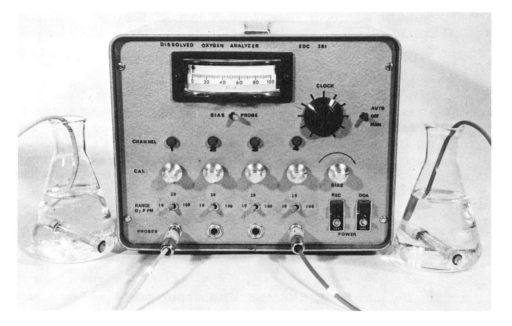

FIGURE 5.11   A four-channel oxygen analyzer. This machine, which is attached to a recording device, registers oxygen concentrations from four different dissolved oxygen electrodes.

Dissolved Oxygen   The measurement of change in concentration of dissolved oxygen in the medium is an accurate measure of bacterial growth. This is applicable to pure or mixed cultures. It has the advantage of being amenable to automation. Figure 5.11 is a photo-graph of a four-channel automatic dissolved oxygen analyzer with four probes reading changes in dissolved oxygen concentration in each of four flasks. The changes in dissolved oxygen are proportional to cell numbers so that they can be used in the study of microbial growth kinetics.

Protein Concentration   Protein concentration is uniform in microbial cells. Measurement of changes in protein concentration can be used to assay growth accurately. This is a particularly useful technique in anaerobic cultures where changes in oxygen concentration cannot be used.

Algae   Growth of unicellular algae can be measured by counting cells under the microscope or by measurement of biomasses. Photosynthetic activity is measured by determining oxygen output or uptake of isotopically labeled $^{14}CO_2$. These methods are discussed in more detail in Chapter 11.

Fungi   Fungal growth is difficult to determine because of the absence of single cells. Colony counts on petri dishes of fungal medium indicate the number of fungal particles and spores but yields a poor estimation of growth rate. Growth is measured by determination of the rate of change in biomasses or of the rate of oxygen uptake in respiratory processes. Protozoan growth is virtually unstudied.

SUMMARY

1. The microbial cell contains carbon, nitrogen, phosphorus, and sulfur in the approximate ratio of 100:10:1:1. Carbon, nitrogen, and phosphorus deficiencies usually limit the extent of microbial growth in natural habitats.

2. Many microorganisms cannot synthesize all their essential growth factors. These are provided in natural habitats by other organisms.

3. Heterotrophic organisms utilize organic energy sources. Chemoautotrophs derive their energy from the oxidation of reduced inorganic compounds. Photoautotrophs utilize the sun's energy.

4. The cell grows by a coupling of catabolic and anabolic reactions. Catabolism may yield energy by respiration or fermentation.

5. Bacterial growth follows four phases: lag, exponential, stationary, and death. Growth kinetics can be developed for either batch or continuous cultures.

6. The growth of bacteria or algae can be determined by a number of different techniques. Fungal growth is usually measured by determination of biomass.

FURTHER READING

H. E. Kubitschek, *Introduction to Research with Continuous Cultures.* Prentice-Hall, Inc., Englewood Cliffs, N.J., 1970.

A. L. Lehninger, *Biochemistry.* Worth, New York, N.Y., 1970.

DEATH
OF
MICROORGANISMS

Death of a microorganism may be defined as the irreversible loss of the capacity to reproduce. The slope of the death curve (Fig. 6.1) indicates the sensitivity of the microbial population to the destructive agent.

Microorganisms are not equally susceptible to killing agents. Each genus and species has specific characteristics that may make it more or less resistant to destruction. The presence of spores substantially improves the survival capacity of a microorganism. The rate of kill is proportional to the effectiveness of the killing agent and the size of the microbial population (Fig. 6.2).

Sterilization is the destruction of all microbial life. This may be achieved by physical means such as heat or ultraviolet light or by chemical means such as ethylene oxide gas. Mixed populations in open ecosystems are rarely sterilized.

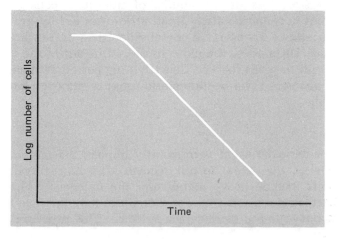

FIGURE 6.1   The end of the stationary and the death phase of bacteria.

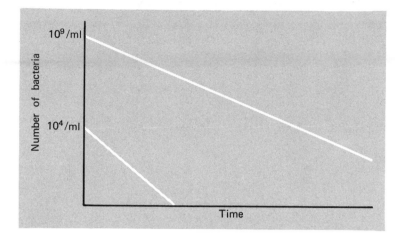

FIGURE 6.2   The effect of cell concentration of the death rate of bacteria. The death rate is much more rapid at low population densities.

Pasteurization is a process in which a sample is heated for a brief period to destroy pathogens. Milk can be pasteurized at 71°C for 15 seconds. Most of the nonpathogenic microorganisms remain viable.

Bacteriostasis   The impairment of an essential metabolic process ultimately results in the death of an organism. When the effect results in a reversible loss of the ability of microbial cells to multiply, the agent is bacteriostatic. Most *antibiotics* act by preventing the synthesis of an essential metabolite, thus preventing cell multiplication. Ultimately, the old cells die, if the antibiotic is kept in contact with the cells for a sufficiently long period of time. The culture will recover if the bacteriostatic agent is removed before the cells all die.

Bactericides   A bactericidal agent permanently impairs the metabolism of the cells so that they do not recover after its removal. Bactericidal agents include heavy metals and the halogens. The addition of a bactericidal agent to a water or waste may kill many microorganisms but will not sterilize the system. This process is called *disinfection*.

PHYSICAL DESTRUCTION

Temperature   The simplest means of destroying microorganisms is to increase the temperature until the cell protein is destroyed. Most living organisms are killed at 100°C. However, the presence of resistant spores in microorganisms requires that higher temperatures be used.

Liquids can be sterilized using an *autoclave*, which is a pressurized chamber in which the sample is subjected to steam at greater than atmospheric pressure. The principle is the same as that used in the pressure cooker. The air in the chamber is replaced by steam. The temperature is maintained at 120°C by a pressure of 15 lb/sq. in. Sterilization is usually achieved in 20 minutes. All microbiological media are sterilized in the autoclave. Boiling water does not kill spores and is, therefore, not sufficiently reliable to be used in laboratory sterilization.

Dry heat is used to sterilize solid materials. The loss of moisture protects spores so that extremely high temperatures are required. A normal sterilization time is 2 hours at 160°C.

Ultraviolet Irradiation   The ultraviolet range of light waves, particularly at wavelengths close to 2600Å, is highly bactericidal. Indeed, the ultraviolet light penetrating our atmosphere from the sun is sufficient to kill unprotected microorganisms in surface waters. The adsorption onto particles serves to protect microorganisms in the air and many of those on surfaces.

Shortwave ultraviolet irradiation is frequently used to sterilize fragile surfaces and rooms. Ultraviolet lamps are installed in hospital operating rooms and in spacecraft manufacturing laboratories. The ultraviolet light has a very low penetrating power so it has not been practicable for water or other liquid treatment. Ultraviolet light is very destructive to the eyes. Even its limited use requires specialized techniques.

Ultraviolet irradiation kills microorganisms by denaturing the nucleic acids. A small fraction of the irradiated cells are *photoreactivated* when they are returned to visible light. The reactivation process appears to involve repair of damaged nucleic acids.

Sonic Oscillation   Electrical generators are available that can produce sonic oscillations in the 5000-10,000 cycles per second frequency range, and ultrasonic oscillation in the range of 200,000 cycles per second. These oscillations, and particularly ultrasonic waves, are very effective for microbial destruction. The high frequencies cause breakage of the cell wall and consequent lysis. The mechanism of sonic destruction is probably *cavitation*. This process involves the production of cavities in the liquid because of rapidly moving bubbles of air. The microbial cell wall, exposed to extreme pressure changes by the movement of these cavities on and off the wall, collapses. Sonic energy is commonly used for cell disruption in research laboratories. It has not been applied successfully for large-scale purification of liquids.

Osmotic Shock Microorganisms maintain high concentrations of materials in their cells in equilibrium with very dilute liquids in the external medium by means of semipermeable membranes. These membranes allow flow of liquids from the dilute medium into the cell. When the medium contains high concentrations of salt, the membrane flow is reversed and the cell is dehydrated (Fig. 6.3). For example, nonmarine bacteria are usually dehydrated and killed in seawater.

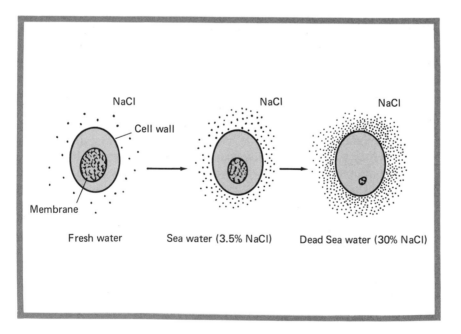

FIGURE 6.3 Dehydration of a freshwater bacterial cell at increasing osmotic concentrations of the external medium. Some dehydration occurs in seawater. Total dehydration occurs in Dead Sea water.

Most microorganisms are killed by high concentrations of salt, which is commonly used in primitive societies as a meat preservative. Very high concentrations of sugar can be substituted for salt with the same result. A few halophilic (salt-loving) bacteria can resist saturated solutions of salt. These can be seen in the Dead Sea or the Great Salt Lake in Utah. Saccharolytic (sugar-loving) bacteria are found in fruit juices. The mechanism of survival of halophilic and saccharolytic microorganisms appears to be related to the capacity to maintain a high internal salt concentration to balance the external osmotic pressure.

Surface Tension   Within the body of a liquid the forces exerted on a molecule are equal (Fig. 6.4). However, molecules at the liquid-air interface are pulled from below and stretch along the interface to produce an elastic film. The force of this interfacial tension is known as *surface tension* and is expressed in dynes per centimeter of surface. The surface tension of a liquid decreases at increasing temperatures.

Reduction of surface tension is associated with surface active agents or *surfactants*. The best known surfactants are soaps and detergents. Surfactants have hydrophilic (water-loving) chemical groups that increase the wetness of a liquid. Soap contains the hydrophilic sodium (Na^+) ion, and detergents the sulfite (SO_3^-) ion.

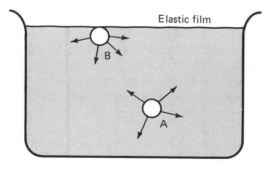

Elastic film

FIGURE 6.4   Surface tension. Molecule A in the body of the liquid is subjected to equal forces on all sides. Molecule B on the surface of the liquid is subjected to attractive forces from below. These molecules on the surface create a surface tension at the air-water interface that forms an elastic film.

The detrimental effect of soaps and detergents on microorganisms results from the increased wettability of the liquid. The surface tension at the cell-liquid interface is reduced, causing either rupture of the cell membrane or toxicity because of cell uptake of the surfactant. Small reductions in surface tension do not adversely affect microorganisms. However, the high concentrations of detergents in domestic use combined with the high temperature of the water is strongly bactericidal.

Filtration   One of the simplest means of removing microorganisms from a liquid is filtration. The most common laboratory filter in use today is the cellulose acetate membrane. Filtration is an important part of the treatment of drinking water. Harmful microorganisms are removed by flocculation followed by passage through sand filters (p. 274). The percolation of partially treated sewage through soil is used to purify sewage in the reclamation water by disposal of sewage onto land surfaces.   During passage of the sewage through the clay profile of the soil, most of the microorganisms are adsorbed onto the clay.   The water reaching the aquifer is almost free of intestinal microorganisms.

DISINFECTION

Chemical disinfection of natural waters has been enormously successful and has saved countless lives during the last half century. The oxidizing chemicals have been most successful.   The most effective chemicals for killing microorganisms in water are

1. chlorine compounds,
2. iodine,
3. ozone,
4. potassium permanganate.

Chlorine and ozone are inexpensive and are the most widely used disinfectants for drinking water. Chlorine compounds are more popular. Ozone is difficult to produce for large-scale use. Because of its efficacy, however, it has been used by some municipalities. Iodine and potassium permanganate have limited use for disinfection of small water supplies.

These oxidizing chemicals are extremely efficient bactericidal agents.   The microbial cell membranes are highly permeable to them.   They act by oxidizing enzymes and metabolites within the cytoplasm.   The nonspecific action of chlorine and other oxidizing disinfectants precludes the possibility of the development of resistant strains of microorganisms.

KINETICS OF DISINFECTION

The rate of disinfection is dependent on the chemistry of the disinfectant, the physiology of the microorganisms, and the environment in which the disinfectant, is acting. The most important factors affecting the rate of disinfection are

1. the nature of the disinfectant,
2. disinfectant concentration,
3. contact time between the disinfectant and the micro-organisms,
4. cell physiology,
5. temperature,
6. organic matter.

Disinfectant   We can obtain a comparative value for a disinfectant by obtaining a K value. This is simply an empirical figure used to compare the efficacy of disinfectants under comparable conditions. To obtain K we apply the equation

$$KC^n t = \log \frac{N_1}{N_2}$$

where C is the disinfectant concentration, n is the dilution coefficient, N_1 is the inoculum concentration, and N_2 is the concentration after disinfection.   It must be emphasized that while K values are useful for comparative purposes, they are highly dependent on the organism being used in the test and on environmental conditions. Unless the K values of two disinfectants have been obtained under identical conditions, their use is seriously impaired.

Concentration   The concentration of disinfectant is critical. Some microorganisms are highly resistant to chemical inactivation, while others are highly susceptible.   Viruses are much more resistant than bacteria to chlorine.

The relationship between disinfectant concentration and the rate of cell destruction is given by the equation

$$C^n t = \text{constant}$$

where C is the disinfectant concentration, N is a constant for the disinfectant, and t is the time necessary to kill a percentage of the microorganisms.

The relative killing effect of hypochlorite (HOCl) on different microorganisms can be seen from this equation. Table 6.1 shows the constants for *E. coli* and three human enteric viruses. Adenovirus is highly sensitive to hypochlorite. However, *E. coli* is much

TABLE 6.1 The disinfectant action of hypochlorite on *E. coli* and three viruses.

Increasing $c^{0.86} t_{99}$ indicates a higher resistance to disinfection.[a]

Organism	$c^{0.86} t_{99}$
Adenovirus 3	0.098
E. coli	0.24
Poliomyelitis virus 1	1.20
Coxsackie virus A2	6.30

[a]From G. Berg, *J. New England Water Works Assn.,* 78:79 (1964).

more sensitive than either poliomyelitis or Coxsackie virus. The concentration of chlorine used to treat sewage or drinking water is based on the ability to kill *E. coli*. It is quite possible that some viruses survive these chlorination levels.

Contact Time Chick's law states that the rate of destruction of microorganisms is proportional to the contact time between the remaining cells and the disinfectant.

$$\frac{dy}{dt} = k(N_1 - y)$$

where y represents $(N_1 - N_2)$ the number of cells remaining, obtained by subtracting the final count (N_2) from the initial count (N_1). k is obtained from the equation

$$\frac{N_2}{N_1} = e^{-kt}$$

Chick's law assumes that all microorganisms in contact with a disinfectant are equally susceptible. This of course is not true. It also assumes that the concentration of disinfectant will not change. The disinfectant concentration usually decreases throughout the contact time, requiring modification of the law.

Physiology The physiological condition of the cells determines their susceptibility to antagonistic chemicals. Young cells are metabolically more active than old cells and are much less resistant to disinfectants. The high rate of permeability of the cell walls and membranes of young cells together with their rapid incorporation of the disinfectant would account for their high rate of susceptibility.

Temperature   The rate of disinfection is increased with increasing temperature. It is relatively easy to determine the temperature coefficient (Q_{10}) for a disinfectant. Q_{10} is the difference in the rate of kill of an organism by a known concentration of a disinfectant between two temperatures ten degrees celsius apart. Table 6.2 provides Q_{10} values for *E. coli* treated with chlorine and chloramine. It is apparent that both chlorine and chloramine activity are strongly dependent on temperature. This is particularly evident at high pH values.

The temperature coefficient is not uniform throughout the temperature range. It may increase with increasing temperature as the organisms become more susceptible or it may decrease if the organisms begin to develop spores or cysts.

Organic Matter   The presence of colloidal organic material decreases the effectiveness of disinfection. The organic matter reduces the activity of the disinfectant by competitively reacting with it. Oxidation of nonmicrobial organic matter puts an additional demand on the disinfectant. The colloids in water may also provide protection for microbial cells against chemical toxicity. The cells may become coated with a protective layer of organic colloids.

TABLE 6.2   The disinfection of *E. coli* with chlorine and chloramine is both temperature and pH dependent.

The Q_{10} represents the increase in reaction rate for a 10°C rise in temperature.[a]

Type of Chlorine	pH	Q_{10}
Aqueous chlorine	7.0	1.65
	8.5	1.42
	9.8	2.13
Chloramine	7.0	2.08
	8.5	2.28
	9.5	3.35

[a]From G. M. Fair, et al., *Water and Wastewater Engineering,* Vol. 2. John Wiley & Sons, Inc., N. Y., 1968.

Viruses are small enough to be protected by adsorption onto colloidal particles.

CHLORINE

Chlorine compounds are inexpensive, easily transported, and very potent antimicrobial agents when dissolved in water. They are almost universally used for disinfection in sewage treatment and for municipal drinking water supplies.

Chlorine is usually added to water or sewage as elemental chlorine gas. The amount required to provide 99.9% destruction of *E. coli* in drinking water should leave a residual 0.2-1.0 mg/liter. The amount required to disinfect sewage is dependent on the concentration of bacteria in the sewage and on the amount of extraneous organic matter present.

Raw sewage may require as much as 24 mg of chlorine per liter to leave a residual of 0.5 mg/liter (Table 6.3). It is not surprising that many maritime municipalities without treatment facilities do not chlorinate sewage going into their ocean outfalls. Treatment lowers the chlorine requirement of sewage. The effectiveness of settling and chemical precipitation of sewage is shown by the reduction in chlorine demand to 3-18 mg/liter. Activated sludge treatment lowers the demand to 3-9 mg/liter. The intermittent sand filter removes most of the organic matter

TABLE 6.3   The approximate amounts of chlorine required to yield a chlorine residual of 0.5 mg/liter after 15 minutes in sewage and sewage effluents.[a]

Type of Sewage or Effluent	Approximate Amount of Chlorine (mg/liter)
Raw sewage	6-24
Settled sewage	3-18
Chemically precipitated sewage	3-12
Trickling filter effluent	3-9
Activated sludge effluent	3-9
Intermittent sand filter	1-6

[a]From G. M. Fair, et al., *Water and Wastewater Engineering,* Vol. 2. John Wiley & Sons, Inc., N. Y., 1968.

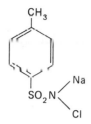

Chloramine-T

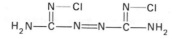

Azochloramide

FIGURE 6.5 Monochloramine and organic chloramines.

from sewage so that only 1-6 mg/liter of chlorine is required to leave a residual of 0.5 mg/liter. Unfortunately intermittent sand filters require large acreage and are rarely used today.

Chlorine gas forms two species, hypochlorous acid (HOCl) and hypochlorite ion (OCl⁻), on contact with water. Both of these species are strong oxidizing agents. The hypochlorous acid dissociates further:

$$HOCl \rightarrow HCl + O$$

The oxygen atom acts as a very strong oxidizing agent reacting with essential cellular components including membranes, nucleic acids, and enzymes to destroy them. The same dissociation occurs with the hypochlorite ion.

Hypochlorite can be obtained as calcium or sodium hypochlorite in liquid form. These compounds are used in domestic bleaches and for disinfection of industrial equipment.

Monochloramine (NH_2Cl) and the organic chloramines (Fig. 6.5) are much more stable than hypochlorite. The release of hypochlorite from monochloramine in water follows the reaction.

$$NH_2Cl + H_2O \rightleftharpoons HOCl + NH_3$$

The chloramines dissociate quite slowly. They are not such effective disinfectants as chlorine or hypochlorite. The slow release of hypochlorous acid makes them particularly useful, however, for single applications at daily intervals. For this reason they are commonly used in swimming pools.

IODINE

Iodine is a good oxidizing agent and is an excellent disinfectant. Its low solubility in water limits its use in water purification. It is a common skin disinfectant, however, when dissolved in alcohol. Iodine is quite stable in water and is much more independent of extraneous organic matter and pH than chlorine. Because of these factors, iodine disinfection of swimming pools is becoming quite common.

Iodine has been used for many years in tablet form for disinfection of small quantities of water in areas where fecal contamination is suspected and no municipal treatment facilities are available. It is particularly useful where cysts of *Entamoeba histolytica* are suspected of contaminating drinking water sources.

OZONE

Ozone is easily produced and is an excellent oxidizing agent. The advantages of ozone are as follows:

1. It has a strong oxidizing ability.
2. Unlike chlorine, it does not produce chlorophenols, which have an unpleasant taste and odor, so it is an effective disinfectant for drinking water containing organic contaminants.
3. It is easily produced at the treatment site. By comparison, chlorine frequently must be carried long distances in pressurized tanks, a dangerous and cumbersome operation.

The major disadvantages of ozone are as follows:

1. It is more expensive to produce than chlorine.
2. Ozone leaves no residual in the water. If there is recontamination following treatment, the fecal organisms are not killed. Chlorine treatment leaves a residual of chlorine in the water that is available to destroy contaminants after the water leaves the treatment plant.

The ozone requirement in typical water treatment plants is 1-1.5 ppm for 10 minutes. The ozone is produced by passing air across a high-voltage electrical discharge. It competes well with chlorine for use in municipal water and wastewater treatment. Many Canadian and European cities now use ozone.

Disinfection rates with ozone are much more rapid than chlorine. It is probable that ozone diffuses through the microbial cell membranes faster. Viricidal activity is also stronger.

PERMANGANATE

Potassium permanganate is not so strong an oxidizing agent as chlorine or iodine. It is therefore not as powerful a disinfectant. It is used in the treatment of drinking water to oxidize organic taste and odor compounds as well as iron and manganese in the water. This treatment destroys some of the microorganisms in the water. Chlorination or ozone treatment is required, however, to ensure a safe drinking water supply.

SUMMARY

1. Sterilization results in a total destruction of microbial populalation. Partial kill is achieved with bacteriostatic chemicals or by pasteurization or disinfection.

2. Physical destruction of microorganisms can be achieved by high temperatures, irradiation, sonic oscillation, osmotic and surface active agents, and filtration.

3. The effectiveness of disinfection is related to cell structure, the nature and concentration of disinfectant, and the contact time. External conditions, such as temperature, pH, or extraneous organic matter, are also important.

4. Chlorine is the most inexpensive and widely used disinfectant for water.

5. Ozone is used in some communities. It has the advantages of being produced on the site and of not producing tastes and odors. It does not provide residual protection, however.

FURTHER READING

G. M. Fair and J. C. Geyer, *Water Supply and Waste Water Disposal*. John Wiley & Sons, Inc., New York, N.Y., 1954. See chapter on disinfection.

G. M. Fair, J. C. Geyer, and D. A. Okun, *Water Purification and Wastewater Treatment and Disposal*. John Wiley & Sons, Inc., New York, N.Y., 1968. See chapter on disinfection.

C. Lamanna and M. F. Mallette, *Basic Bacteriology*, 3rd ed. Williams & Wilkins Co., Baltimore, Md., 1965. See chapters on physical factors affecting bacteria and on disinfection.

W. J. Weber and H. S. Posselt, "Disinfection," in *Physical Processes for Water Quality Control* by W. J. Weber. John Wiley & Sons, Inc., New York, N.Y., 1972.

WATERBORNE

PATHOGENS

7

WATERBORNE PATHOGENS

In those parts of the world where community drinking water is disinfected, the incidence of waterborne disease, with some exceptions, is very low. Since the turn of the century the mortality from typhoid and paratyphoid in the United States has declined from 37 per 100,000 to less than 1 per 100,000 (Fig. 7.1). The potential for drinking water contamination is always present. In many parts of the world, drinking water is untreated and contaminated with pathogens. Some waterborne pathogens are not transmitted in drinking water. Schistosomiasis is contracted by adsorption of the pathogen through the skin. Hepatitis and occasionally typhoid may be caused by ingestion of shellfish grown in polluted waters.

BACTERIA Three genera comprise the most important group of bacteria potentially capable of contaminating drinking water supplies and causing disease in man. These are

Salmonella,

Shigella,

Vibrio comma.

Salmonella Contamination of drinking water with *Salmonella* may cause outbreaks of gastroenteritis or typhoid. The source of the bacteria causing gasteroenteritis may be either human or

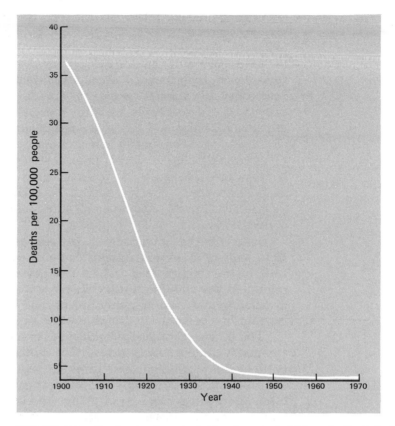

FIGURE 7.1 Death rates from typhoid and paratyphoid fever in the United States between 1900 and 1970. The mortality has declined from 37 per 100,000 in 1900 to less than 1 per 100,000 in 1970. (From M. J. Pelczar, Jr. and R. D. Reid, *Microbiology,* 3rd ed. McGraw-Hill, Co., N. Y., 1972.)

animal feces. Typhoid is caused by *Salmonella typhosa*, which is only transmitted in human feces.

Salmonella is a motile Gram-negative rod that does not produce spores. It is aerobic and grows well on most nutrient media. The temperature optimum for growth is 37°C, which is typical for human pathogens. *S. typhosa* ferments glucose, producing acid but no gas.

Severe gastroenteritis resulting in acute diarrhea and vomiting may also be caused by *S. paratyphi*, which is transmitted only by humans, or by other Salmonellae carried by animals. Examples include *S. chleoraesuis* carried by swine, *S. typhimurium* carried by rodents, and *S. enteritidis*, which is common in fowl and other domestic animals.

Typhoid is a much more serious disease in which the intestine is inflamed and the spleen is enlarged. Severe toxic effects occur and the disease is accompanied by a high temperature. The organisms are excreted in feces and urine in large numbers during the active phase of the disease and for a short period after symptoms have disappeared. Rare individuals do not display any symptoms of disease; yet they are chronic excretors of the pathogen. These people are known as *carriers*. They frequently initiate epidemics and are often responsible for the maintenance of a reservoir of *Salmonella* in a community.

Salmonella has a short survival time in natural waters at temperatures above 15°C. Within 7 days most Salmonellae have been destroyed. The mechanism of destruction is discussed in Chapter 13. In very cold soil and water *Salmonella* can survive for periods of years.

Prior to the introduction of sewage and drinking water treatment, typhoid was a major cause of death in western Europe and North America (Fig. 7.1). The development of large urban centers in the midnineteenth century was quickly followed by the installation of sewage treatment plants and by the provision of safe typhoid-free community drinking water supplies.

Today waterborne epidemics of typhoid in the developed countries are extremely rare. The potential for epidemics of typhoid from contaminated drinking water is high, however, in the developing countries. Modernization and urbanization are often proceeding more rapidly than the installation of safe community drinking water supplies. The rapid development of high population densities without stringent control of sewage disposal and drinking water supplies provides a serious danger of waterborne *Salmonella* infections.

In the absence of adequate sanitation, the community can be protected by immunization. Inoculation with killed suspensions of *S. typhosa*, together with *S. paratyphi*, gives immunity against typhoid and gastroenteritis caused by *Salmonella*.

Shigella Contamination of drinking water or food with feces containing *Shigella* causes epidemics of bacillary dysentery. Human feces is the major source of the bacterium. *Shigella* is a nonmotile Gram-negative rod that does not form spores. It can be differentiated from *Salmonella* by the absence of motility and by the inability of *Shigella* to grow on gelatin or to produce hydrogen sulfide. *Shigella* is aerobic and grows well on nutrient media at 37°C. Neither *Salmonella* nor *Shigella* ferment lactose. This characteristic is used to differentiate them from *E. coli*.

Shigella is excreted in feces and urine during the active phase of the disease. The organism does not grow in natural waters and rarely survives more than 10 days. Bacillary dysentery is characterized by diarrhea. It is rarely fatal. The number of notified cases has steadily increased during the past 60 years. Figure 7.2 illustrates graphically the apparent rise in the incidence of bacillary dysentery in England and Wales. However, the increase is the result of more efficient notification procedures. The decline in the number of deaths in the United States caused by *Shigella* during the past 50 years parallels the decline in typhoid mortality (Fig. 7.1). The decrease in mortality reflects the trend toward prevention of spread of *Shigella* because of the development of modern sanitation techniques. The control of *Shigella* and *Salmonella* was further accelerated by the development of antibiotics that effectively control typhoid and paratyphoid and prevent the spread of epidemics.

Waterborne epidemics of bacillary dysentery caused by *Shigella* occasionally occur in developed countries. Outbreaks usually are

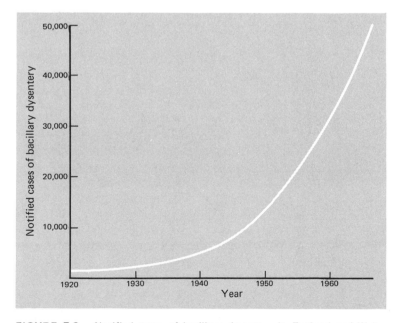

FIGURE 7.2 Notified cases of bacillary dysentary in England and Wales between 1920 and 1960. (After G. S. Wilson and A. A. Miles, *Principles of Bacteriology and Immunology,* 5th ed. Edward Arnold Ltd., London, England, 1964.)

in small communities where fecal contamination of untreated water supplies occurs. In developing countries waterborne bacillary dysentery is common. No means of artificial immunization has yet been developed. Eradication is dependent on adequate hygiene to prevent fecal contamination of food and water.

Vibrio comma   Cholera is caused by a short curved Gram-negative rod, *Vibrio comma*. The bacterium is motile and does not form spores. It is aerobic and grows on nutrient media at 37°C. Cells of *V. comma* are shown in Fig. 7.3. Cholera is a severe gastrointestinal disease of humans and is caused by eating food or drinking water contaminated by feces containing the vibrio. A toxin is liberated in the intestine causing vomiting and severe diarrhea. The disease has a high mortality rate caused by rapid dehydration and loss of minerals. Treatment involves replacement of liquids and minerals, together with antibiotic therapy.

Cholera epidemics have decimated whole communities by the rapidity of spread and the high fatality rate. Large numbers of

FIGURE 7.3   *Vibrio comma,* the causative organism of cholera. The bacteria divide to form comma-like cells.

TABLE 7.1   Deaths from cholera in India and East Pakistan in 1956.
The mortality rate is very high.

Country	Population	Deaths	Mortality per 100,000
India	377 million	24,000	6.8
East Pakistan	42 million	18,000	4.4

the pathogen are excreted during the course of the disease and for a few days afterward. Cholera vibrios are quite sensitive to adverse conditions and rarely survive more than 7 days in natural waters. In periods of an epidemic, however, there is frequently continuous contamination of drinking water supplies with *V. comma*.

Cholera is *endemic* in Asia. There is a reservoir of the bacterium present in the population that explodes into epidemics when a large segment of the population is exposed to the bacterium in food or drinking water. Typical mortality figures for some Asian countries are summarized in Table 7.1. Fatality varies from 5 to 75% depending on the severity of the infection and on the availability of medical treatment. In 1947 a cholera epidemic in Egypt killed more than 20,000 people.

Cholera is very rare in developed countries. This can be attributed directly to the high level of hygiene and, in particular, to the destruction of pathogens in sewage treatment plants and by the disinfection of drinking water. Immunization against the disease can be obtained by inoculation with dead cells of *V. comma*. Immunity lasts approximately 6 months.

VIRUSES

Infectious Hepatitis   The virus that causes infectious hepatitis is the only documented waterborne viral pathogen. Numerous outbreaks of infectious hepatitis have been traced to fecal contamination of either drinking water supplies or shellfish beds.

Very little is known about the virus. It is extremely difficult to cultivate in tissue culture. Identification of the pathogen is achieved by typical symptoms. The disease is characterized by a yellow jaundice of the skin caused by an enlarged liver, vomiting, and abdominal pain. It may last as long as 4 weeks. Mortality is low.

Explosive outbreaks of infectious hepatitis are common, particularly where water treatment has failed. An epidemic in New Delhi, India, in 1955 was traced to fecal contamination of the water supply. Between 20,000 and 200,000 cases of infectious hepatitis occurred. The virus is excreted in feces and urine during and directly after the active phase of the disease.

The hepatitis virus is endemic throughout the world. At least 60,000 people contract the disease in the United States each year. Most of these cases originate in shellfish contaminated with feces containing the virus.

Control of infectious hepatitis is dependent on adequate treatment of sewage and water. Viruses are more resistant to chlorination than bacteria. Some outbreaks of infectious hepatitis have been attributed to inadequate chlorination of domestic water supplies. This question has not been resolved and requires further research. No artificial immunization is available.

Enteroviruses   These viruses belong to the *picorna* group. They are very small, less than 25 mμ in diameter, and their nucleic acid is ribonucleic acid. Hence their name, pico (small), and RNA: picorna. The enteroviruses are a subgroup of the picorna viruses that live in the intestine. They include polio virus, which causes poliomyelitis; Coxsackie viruses; and echo viruses.

There is no strong evidence that poliomyelitis virus is transmitted by water. Coxsackie and echo viruses are very common in human feces, however, and persist in untreated water for long periods of time. They may be of importance as the cause of mild intestinal disorders.

PROTOZOA   The amoeba *Entamoeba histolytica* is the causative organism of amebiasis, also known as amebic dysentery. The disease is contracted by eating food or drinking water contaminated by feces containing the pathogen. The disease may manifest itself as either a mild diarrhea or chronic dysentery.

The amoeba produces cysts that can survive as long as 6 months in natural waters. They are not resistant to chlorination. Amebiasis is eliminated in communities using chemically treated drinking water.

Recently, another amoeba, *Naegleria gruberi*, has been shown to cause a fatal meningoencephalitis in people swimming in certain lakes and ponds. A small number of cases have been reported from Florida, Australia, and Czechoslovakia (Table 7.2). Apparently a reservoir of the amoeba is present in certain ponds. Less

TABLE 7.2 Known outbreaks of amebic meningoencephalitis caused by the protozoan *Naegleria grubeii.*[a]

Location	Number of Cases
Orlando, Florida	4
South Australia	6
Richmond, Virginia	12
Czechoslovakia	17

[a]After E. E. Geldreich, *Water Pollution Microbiology,* ed. R. Mitchell. John Wiley & Sons, Inc., N.Y. 1972.

than 100 cases have been reported, most of them from three or four distinct ponds. The existence of reservoirs in nature of a fatal pathogen in natural waters has stimulated research into the possibility that *Naegleria gruberi* has an alternate host to man.

SCHISTOSOMIASIS

The parasitic *trematodes* or *blood flukes*, although not strictly microorganisms, are by far the most important waterborne human pathogens. These organisms are tiny worms between 15 and 20 mm large. The genus *Schistosma* is responsible for the debilitating disease schistosomiasis. A conservative estimate of the number of people suffering from the disease is 114 million. Three species attack humans:

1. *Schistosoma japonicum* occurs in the waters of Japan, Korea, the Phillipines, and China. The frequency of infection is from 10 to 25% of the population.

2. *S. hematobium* is prevalent in East Africa. In some areas more than 50% of the population are infected.

3. *S. mansoni* is predominant in Africa, Egypt, and South America.

The symptoms of schistosomiasis are enlargement of the liver, diarrhea, and anemia. The disease is chronic and leads to intestinal ulceration. Excretion of the eggs in feces or urine into natural waters frequently leads to reinfection and serves to spread the disease.

Life Cycle The life cycle of *Schistosoma* is illustrated in Fig. 7.4. The female worms are larger and narrower than the males and lie

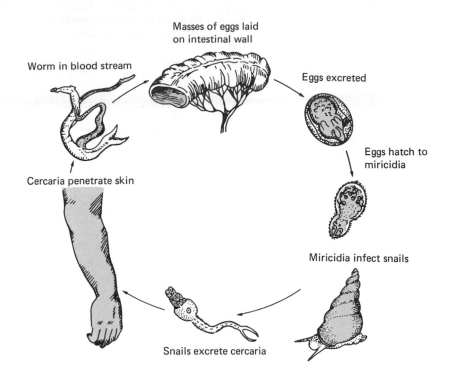

FIGURE 7.4 The life cycle of *Schistosoma,* the causative agent of schistosomiasis. (After R. D. Barnes, *Invertebrate Zoology,* 3rd ed. W. B. Saunders Co., Philadelphia, Pa., 1974.)

within a groove along the length of the male. In human hosts they live in the veins and lay eggs on the intestinal walls. Each worm may excrete thousands of eggs each day.

The eggs hatch in water to yield motile miricidia. Each miricidium can only live for a few hours. If it cannot find a suitable snail host, it dies. The miricidia that infect snails undergo morphogenesis in a month to yield another motile form, the *cercaria.* This cell is about 150μ long and has a forked tail. The cercariae have a life span in the water of 2 or 3 days. Huge numbers are excreted by the snail so that there is ample opportunity for human infection. The cercariae attach to the skin by suckers. They produce an enzyme hyaluronidase, which dissolves body tissue and allows them to enter the blood stream. They flow through the circulatory system to the liver where they mature to adult trematodes.

Control is difficult because of the lack of adequate sanitation in endemic areas. The eggs are being continuously excreted into water used for bathing, washing, and drinking. With the development of modern irrigation methods the disease is spreading at a much more rapid pace in the irrigation waters. The use of molluscicides to control the snail host is partially effective. No adequate immunization or mass chemotherapeutic agent has been developed. The ultimate control of schistosomiasis must await the development of modern water and sewage treatment facilities in those areas of the world where it is endemic.

DETECTION OF FECAL CONTAMINATION

Escherichia coli is a prominent member of the community of microorganisms living in the human intestine. Large numbers are found in human feces. For this reason the detection of *E. coli* in natural waters is used as an indication of fecal contamination of water. This does not confirm the presence of pathogenic microorganisms in the water but rather suggests the possibility of their presence.

The laboratory tests for fecal contamination begin with a broad estimation of the probable number of *coliform* bacteria in the sample. Coliforms include *Salmonella*, *Shigella*, and *Escherichia coli*. In addition, the common soil bacterium *Aerobacter* is a coliform and may provide false evidence of fecal contamination. The tests for fecal contamination are carried out in three stages (Fig. 7.5):

1. presumptive test,
2. confirmed test,
3. completed test.

Presumptive Test   The presumptive test involves the inoculation of water samples into test tubes containing lactose broth. Small inverted test tubes in the medium catch gas produced during the growth of the bacterium. The production of gas provides presumptive evidence of fecal contamination.

Confirmed Test   The confirmed test for fecal contamination is based on the subculture of bacteria from positive tubes in the presumptive test to petri dishes containing a nutrient medium and the dyes eosin and methylene blue. *Escherichia coli* colonies have a metallic green sheen on eosin methylene blue (EMB) agar (p. 30). *Aerobacter aerogenes* colonies are large and pink on EMB agar.

Completed Test The test for fecal contamination is completed by transferring colonies from EMB agar to a liquid medium containing lactose. The production of gas following 24 hours incubation at 37°C by typical nonsporing Gram-negative rods provides confirmation of fecal contamination.

FIGURE 7.5 A scheme for the enumeration of enteric bacteria in water and wastewater. Three steps are involved. (a) Presumptive test. (b) Confirmed test. (c) Completed test. All incubations are at 37°C.

(a) Presumptive test

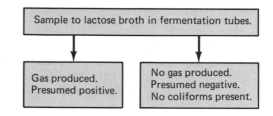

(b) Confirmed test

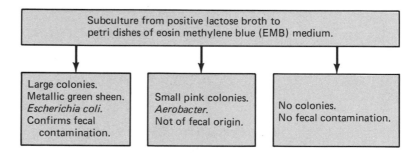

(c) Completed test

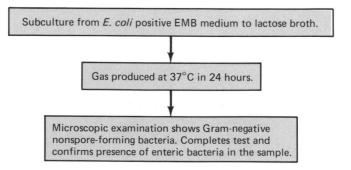

An alternative method of testing for fecal contamination that is frequently used in Europe involves dilution of the water sample in tubes of lactose broth and incubation at 44°C. *E. coli* grows at this temperature, whereas *Aerobacter* is inhibited. This technique, known as the Eijkman test, differentiates fecal coliforms in one step.

Another modification of the coliform test is becoming common for routine analysis of drinking waters or other water samples where the expected coliform count is very low. In this test, known as the membrane filter technique, the water is filtered through a bacteriological membrane that holds back the bacteria. The membrane is incubated directly on lactose eosin methylene blue agar. Colonies of *E. coli* are counted directly on the membrane following incubation (Fig. 7.6).

DETECTION OF VIRUSES

Enteric viruses have been found in water treated to eradicate coliform bacteria. Tests are being developed to specifically assay for enteric viruses. Emphasis is placed on drinking water supplies, where even a small number of virus particles would pose a threat to the community.

Few laboratories have the capability of testing water routinely for contamination with enteric viruses. Since viruses are incapable of growing outside the host cell, either whole animals or tissue culture must be used to grow the organisms. For the cultivation of enteric viruses, monkey kidney cells are usually used. The tissue culture is inoculated with a sample of water. The concentration of virus particles in the water is estimated from the number of plaques observed for each milliliter of water sample.

Detection of enteric viruses in water is hampered by their low concentration. Frequently as few as 1 virus unit per liter is detected. Routine analysis of natural waters may yield 1 positive sample out of 10. This problem is overcome by techniques that concentrate viruses from large quantities of water. A number of methods have been developed. These include

1. gauze pad,

2. ultra-filtration,

3. Polymer treatment.

Gauze pad   The gauze pad technique has been used for many years. Pads are suspended in the sewage or water for 24-28 hours. The

(a)

(b)

FIGURE 7.6 The use of Millipore filters for analysis of water. (a) A Millipore filter is placed on the filter holder. (b) A filter funnel is locked onto the filter. (c) A measured quantity of water to be tested is filtered under negative pressure. (d) The filter is removed and placed on an eosin methylene blue medium in a petri dish. (e) Following incubation at 37°C for 24 hours, the number of typical *E. coli* colonies on the filter is counted to give an estimate of the coliform count in the water. (Courtesy Millipore Filter Corp.)

126

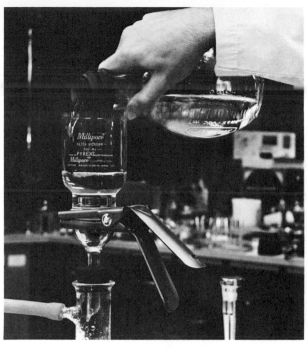

(c)

(d)

(e)

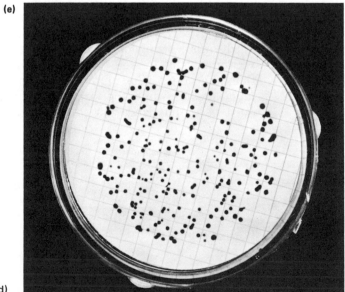

FIGURE 7.6   (continued)

127

expressed liquid contains 10 to 50 times the virus concentration of the water.

Ultracentrifugation Ultracentrifugation is most commonly used for virus concentration. Centrifugation is an ultracentrifuge at 60,000 X gravity for 1 hour is required to deposit the virus particles. Continuous ultracentrifugation can concentrate more than 2 liters of water per hour. Ultracentrifugation provides the high level of concentration necessary for assay of viruses in water. A serious limitation is the expense of the equipment.

Polymer treatment The addition of polymers can be used to concentrate viruses in water samples. A combination of dextran sulfate and polyethylene glycol added to the water forms a separate phase in which the viruses concentrate. A concentration of 500 can be obtained by this method.

A wide range of materials can be added to water to absorb viruses. These include polyelectrolytes and ion-exchange resins. The absorbants containing the viruses are separated from the large volume of water by centrifugation. This concentration technique is inexpensive and highly efficient.

STANDARDS

The numbers of coliforms permitted in water is dependent on the purpose for which the water is to be used and the population density it serves. The standards for drinking water obviously are much more stringent than those for recreational waters. The coliform count in well water serving a single family can be higher than the count in a municipal water supply since the danger is much greater when the size of the population served is large. It must be emphasized that standards set using coliform counts are

TABLE 7.3   Water standards for the United States.

Type of Water	Maximum Permissible Coliform Count (per 100 ml)
Drinking water	1
Recreational waters	1000
Waters used for shellfishing	70

arbitrary. They represent a measure of degree of *possible* human fecal contamination. The potential presence of pathogens is only a statistical possibility related to the level of fecal contamination and the presence of pathogens in the water. The standards used for water in the United States are summarized in Table 7.3.

Drinking Water   One coliform per 100 ml is the maximum count allowed in municipal drinking water supplies. It permits a statistical possibility of contracting disease from the presence of a very small quantity of fecal material. In practice the figure of 1 per 100 ml seems justified since disease has not been associated with drinking water used with these standards.

Recreational Waters   The standards for recreational water are much more difficult to determine. Usually the coliform count is not allowed to exceed 1000 per 100 ml of water in bathing waters. This is an entirely arbitrary figure. Studies in Great Britain have shown that there is virtually no risk of intestinal disease from swimming in sewage contaminated seawater. Despite this fact, public health officials close bathing beaches on the basis of high coliform counts. We have no detailed information on the risks from swimming in contaminated fresh waters.

Swimming Pools   The criteria for swimming pools are quite different. *Pseudomonas aeruginosa* causes ear, nose, and throat infections in inadequately treated pools. The source of *Pseudomonas* may be fecal or it may come from skin infections. There is, in addition, little resemblance between the survival curves of *E. coli* and *Pseudomonas* in natural waters. The coliform standard is of minimum use in the determination of potential *Pseudomonas* infections.

Shellfish   The U.S. Public Health Service standards approve shellfishing in waters in which the coliform count is below 70 per 100 ml. The collection of shellfish is prohibited when the coliform count exceeds 2300 per 100 ml; this indicates gross fecal contamination. Between these two figures, fishing is approved on a restricted basis if it can be demonstrated that the high count is not of fecal origin, e.g., from agricultural wastes. Shellfish provide a reservoir for infectious hepatitis. Therefore, determination of contamination with human feces is of prime importance in shellfish beds.

SUMMARY

1. The important waterborne bacterial pathogens are *Salmonella*, *Shigella*, and *Vibrio comma*. Protozoan pathogens include *Entamoeba histolytica* and *Naegleria gruberi*. Infectious hepatitis is the most important viral pathogen.

2. The trematode *Schistosoma* is the agent of schistosomiasis. The organism has a complex life cycle. Control depends on achieving a break in the life cycle.

3. Fecal contamination of water is indicated by detection of *Escherichia coli*. Standards for drinking water, recreational water, and shellfish beds are based on the coliform count.

4. Enteric viruses are not routinely assayed in natural water. They pose a threat to drinking water supplies. Methods are being developed to concentrate viruses from large quantities of water.

5. The maximum coliform count permissible for drinking water in the United States is 1 per 100 ml of water. Usually 1000 coliforms per 100 ml are permitted in bathing waters. The collection of shellfish is prohibited when the coliform count exceeds 2300 per 100 ml.

FURTHER READING

"Water and Man's Health," U.S. Agency for International Development, Washington, D.C., 1962.

G. Berg, *Transmission of Viruses by the Water Route.* Wiley-Interscience, New York, N.Y., 1967.

E. E. Geldreich, "Water-Borne Pathogens" in *Water Pollution Microbiology*, ed. R. Mitchell. John Wiley & Sons, Inc., New York, N.Y. 1972.

Standard Methods for the Examination of Water and Wastewater, 13th ed. American Public Health Association, Washington, D.C., 1971. See section on bacteriologic examinations.

G. S. Wilson and A. A. Miles, *Principles of Bacteriology and Immunology*, 5th ed. Edward Arnold, Ltd., London, England, 1964.

ORGANIC MATTER

DECOMPOSITION

8

ORGANIC MATTER DECOMPOSITION

There is a continuous influx of carbonaceous material to soil and water. The *turnover* or *biodegradation* of this material is fundamental to life on earth.

Biodegradation of carbon compounds is essential for two reasons:

1. All living organisms have carbon as their chief building block. In the absence of degradation of dead organisms and organic wastes, we would be inundated with the remains of animals, plants, and microorganisms.

2. Conversely, biodegradation releases essential nutrients for the growth of other organisms. This mineralization of organic material is necessary to maintain the pool of nutrients for biological activity without rapidly draining the reserves in the earth's mantle.

SOURCES

Three sources of carbonaceous compounds are available to microorganisms for biodegradation:

1. dead organisms,

2. excreta,

3. nonbiogenic carbon compounds.

All organisms, from animals to protists, are ultimately biodegraded and their nutrients returned to the abiotic pool in soil and water.

Dead organisms do not usually upset the biological equilibrium. They are decomposed in the soil. A major source of disturbance comes from human and animal excreta in our natural waters. In urban areas a serious threat to our natural waters comes from the large quantity of human excreta. Sewage treatment plants provide a microflora to degrade the carbon compounds that the native microflora in natural waters cannot handle because of the great quantities.

In rural areas, however, the quantities of animal excreta far exceed the amount of human feces in urban areas. A major factor contributing to the problem of disposal of domestic animal wastes is the *urbanization of farm animals*. Livestock in the United States produce more than 1 billion tons of solid wastes and almost ½ billion tons of liquid wastes annually (Table 8.1). The amount of organic matter is equivalent to that produced by 2 billion people. A typical feedlot for cattle contains 10,000 animals and is equivalent in organic matter load to a city of 45,000 people. About 11 million cattle are on feedlots at any one time.

TABLE 8.1 Huge quantities of wastes are produced by livestock.

A significant amount of these wastes are concentrated because of the urbanization of farm animals. About 10% of all cattle in the United States are held in feedlots.

Livestock	*Population (millions)*	*Annual Production of Solid Wastes (million tons)*	*Annual Production of Liquid Wastes (million tons)*
Cattle	107	1004.0	390.0
Horses	3	17.5	4.4
Hogs	53	57.3	33.9
Sheep	26	11.8	7.1
Chickens	375	27.4	—
Turkeys	104	19.0	—
Ducks	11	1.6	—
Totals		1138.6	435.4

The third group of carbonaceous compounds being disposed into soil and water is nonbiogenic. These include fuel, oils, pesiticides, and industrial wastes. Many of these materials are *recalcitrant*, i.e., difficult to degrade biologically. These recalcitrant materials are considered separately in Chapter 9.

The quantities of carbonaceous wastes produced by households and industrial plants in the United States are extraordinary; 120 million people are served by sewers. These people produced 5,000 billion gallons of wastewater with a combined biochemical oxygen demand (B.O.D.) of more than 7 billion lb in 1963 (Table 8.2). The manufacturing industries produced 22 billion lb of B.O.D. (B.O.D. is discussed in detail on page 138.) The chemical, food, and paper industries are responsible for most of the industrial wastes (Table 8.2).

TABLE 8.2 Domestic and industrial sources of organic pollutants entering natural waters in the United States for 1963.

Source	Waste Water Volume (billion gal/yr)	Strength (B.O.D.) (million lb/yr)
Domestic wastes	5300	7300
Chemical industries	3700	9700
Pulp and paper industries	1900	5900
Textile mills	140	890
Food Processing	690	4300
Petroleum and coal	1300	500
Rubber and plastics	160	40

THE CARBON CYCLE   The prime source of carbon for living organisms is the carbon dioxide in the air. The transfer of carbon between the abiotic and biotic environment is shown in Fig. 8.1.

Carbon dioxide from the air is fixed into biological tissue by the primary-producing photoautotrophs, the plants and algae. Chemoautotrophs also fix atmospheric carbon dioxide. However, their importance as a means of capturing carbon for biological processes is far less than the photoautotrophs.

The food web is the vehicle for carbon to be cycled through

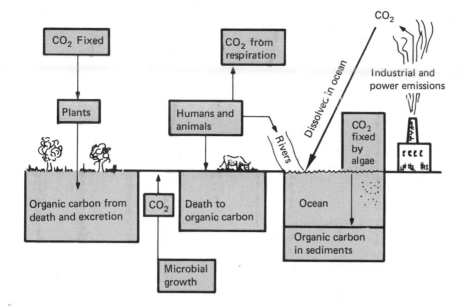

FIGURE 8.1   Transfer of carbon between CO_2 and organic carbon in the atmosphere, in oceans, and on land.

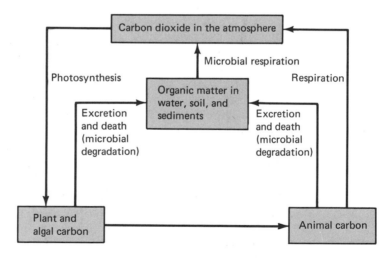

FIGURE 8.2   The carbon cycle. In aquatic ecosystems the CO_2 in the atmosphere is fixed by photoautotrophs. Organic carbon moves through the food chain. After death it is either recycled by the microflora or laid down in the ocean sediments.

living organisms (Fig. 8.2).   Heterotrophs consume autotrophs, utilizing their carbon as a substrate for growth and energy. Heterotrophic microorganisms obtain their substrates for growth and energy by degrading carbon compounds in plants, algae, animals, and other microorganisms. Organic excreta are also utilized by microorganisms as a carbon source.

Nonliving organic matter plays an important part in the carbon recycling process (Fig. 8.1). There is a pool of organic compounds in soil and water produced by the microbial degradation of dead organisms and by excretion products. This organic matter is the source of nutrients for large populations of heterotrophic microorganisms living in natural habitats. In addition, the pool is utilized as a store for inorganic nutrients required for plant growth as well as a source of essential growth factors that plants and algae are incapable of synthesizing. A significant portion of the carbon dioxide utilized in biological processes is returned to the air. Plants respire in the dark, releasing carbon dioxide. All animals respire, continually returning carbon dioxide to the air. Microorganisms, in their continual turnover of organic matter, convert part of the organic substrate to cells, part to other organic extracellular metabolites, and a portion to carbon dioxide.

AEROBIC DECOMPOSITION

A wide range of organic compounds continuously bombard the microbial environment. There is an equally complex system of microorganisms present in the ecosystem producing enzymes that degrade these substrates. The indigenous microflora is in a continuous state of flux as it adapts to changing substrates. The microflora that predominates in the presence of one substrate may be almost totally replaced when another substrate enters the habitat.

The rate of decomposition is also tied to the concentration of other elements in the system. The ratio of carbon:nitrogen:phosphorus in the bacterial cell is approximately 100:10:1. There is a direct proportionality between the rate of decomposition of organic substrates and the concentration of nitrogen and phosphorus in the ecosystem. This relationship is only valid at C:N:P ratios below 100:10:1.

Decomposition continues until some essential element becomes limiting. In the turnover of algae in lakes and in the sea, nitrogen usually becomes deficient and prevents further decomposition. Figure 8.3 shows how the C:N ratio of the substrate changes as the microorganisms assimilate the carbon. When the C:N ratio reaches approximately 10 and the C:P reaches 100, the turnover stops. The C:N:P for Atlantic Ocean water is maintained at a constant 100:10:1 (Fig. 8.4). Sewage treatment plants are nitrogen deficient. Addition of nitrogen would increase the efficiency of secondary treatment but would cause a large eutroph-

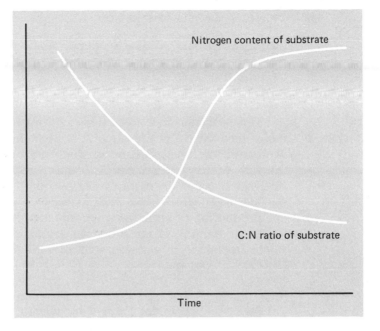

FIGURE 8.3 The relationship between C:N ratio of a substrate and the nitrogen content of the remaining substrate during organic matter decomposition.

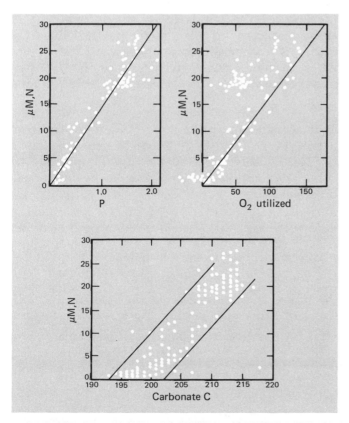

FIGURE 8.4 The stoichometric correlation among carbonate carbon, soluble nitrate, phosphate, and oxygen utilized in the western Atlantic. The concentrations are in millimicrons. It is apparent that there is a close correlation between the utilization of nitrogen and phosphorus as a result of photosynthetic activity. The ratio of carbon:nitrogen:phosphorus in the ocean approximated 106:16:1, which mirrors microbial cell composition. (From W. Stumm and E. Stumm-Zollinger, *Chimia,* 22:325 (1968).)

ication potential. Similarly, paper waste being disposed of into rivers could be more rapidly degraded by the addition of nitrogen.

**BIOCHEMICAL
OXYGEN DEMAND**

The aerobic decomposition of organic substrates by the indigenous microflora in aquatic ecosystems is accompanied by a decline in the concentration of dissolved oxygen in the water. If the rate of oxygen utilization is more rapid than the reaeration rate in the water, then the water becomes anaerobic. The consequences of anaerobiosis include a decline in the fish population and the production of unpleasant volatile metabolic products.

The potential rate of substrate degradation and oxygen utilization is assessed by measurement of biochemical oxygen demand (B.O.D.) of the water.

The Test   B.O.D. is determined by placing the sample in a sealed bottle. The material is incubated in the dark at $20°C$ for 5 days. The amount of organic matter available for biodegradation at the beginning is measured at the end of the incubation period by determining the amount of oxygen consumed. The Winkler test for dissolved oxygen is used to measure the amount of oxygen remaining in the bottle. The test is based on the oxidation of manganous ion by dissolved oxygen. Iodide is added to the sample and reduces the manganic ions, liberating iodine. The liberated iodine is titrated against sodium thiosulfate. Alternatively, oxygen can be measured directly using oxygen electrodes.

B.O.D. Load   Engineers normally utilize the B.O.D. test to determine the depletion of oxygen by microorganisms growing in natural waters. The addition of available organic substrate increases the B.O.D. The load of organic matter can be determined from the first-order equation

$$y = L(1 - e^{kt})$$

L = first-stage B.O.D.

y = oxygen demand exerted in time t

r = a constant for the test, usually 0.39

It should be noted that L represents the first-stage B.O.D. Two further stages of respiration occur. These are illustrated in Fig. 8.5.

In the first stage the heterotrophic microflora utilize the avail-

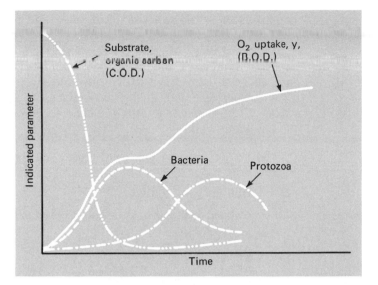

FIGURE 8.5 Microbial activity in a B.O.D. bottle. As the organic substrate declines, oxygen uptake increases. The primary microbial population is bacterial. As the substrate declines, a secondary population of protozoa develops. The protozoa utilize the bacteria as a substrate. (From A. Gaudy in *Water Pollution Microbiology,* ed. R. Mitchell. John Wiley & Sons, Inc., N. Y., 1972.)

able nonrecalcitrant organic compounds in the water. During growth they assimilate the organic matter and excrete reduced inorganic metabolites. At the end of the first stage all available organic compounds have disappeared and the medium is rich in ammonium compounds. In the second stage, autotrophic bacteria that oxidize ammonium to nitrate proliferate, utilizing CO_2 as a carbon source.

The third stage of oxygen utilization involves predation of protozoa and other predaceous microorganisms on the large bacterial population developed in the two earlier stages.

The five day B.O.D. test measures the first stage heterotrophic microflora. The two later stages are ignored. When an engineer speaks of a B.O.D. of 50 mg/liter or of removal of 99% of the B.O.D. in a waste by some treatment process, he is referring only to the immediately available organic matter.

The Streeter-Phelps Equation Fast-running streams, the upper layer (or *epilimnion*) of lakes, and the surface waters of the open ocean are usually oxygen saturated. Slowly moving streams, the cold deep water (or *hypolimnion*) of lakes, and the depths of the ocean are commonly oxygen deficient.

The ability of microorganisms in a stream to degrade a specific concentration of organic waste can be deduced by observing the rate of oxygen utilization and the rate of reaeration from the atmosphere and algal photosynthesis. This relationship is the basis of the Streeter-Phelps equation to describe the oxygen deficit in a natural water:

$$D = \frac{La}{f-1} e^{-rt} \left\{ 1 - e^{-(f-1)kt} [1 - (f-1)\frac{Da}{La}] \right\}$$

D = final oxygen deficit

La = oxygen utilized in substrate degradation

t = elapsed time

Da = initial oxygen deficit

f = rate of self-purification

f = R/r, where R = rate of reaeration and r = rate of deoxygenation

Dissolved oxygen sag curves (Fig. 8.6) are obtained using the Streeter-Phelps equation. The oxygen sag curve allows the engi-

FIGURE 8.6   The oxygen sag curve of Streeter and Phelps. The oxygen concentration of the stream declines as the microflora utilizes an organic pollutant. Heterotrophic activity declines as the substrate disappears. Reaeration is by atmospheric oxygen and photosynthesis. The diurnal oxygen curve demonstrates the importance of the algae in the reaeration process.

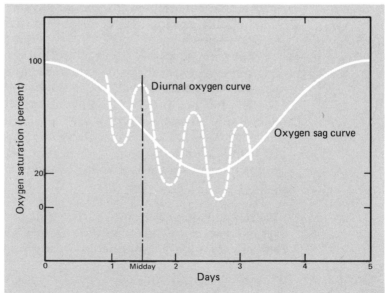

neer to integrate the load of available substrate added to a natural water with the amount of dissolved oxygen in the water and the ability of the water to reaerate and recover from a depletion in its oxygen supply. A fast-flowing stream, because of its rapid reaeration rate, can assimilate a much higher quantity of organic substrate than a stagnant stream. The oxygen concentration is diurnal, reflecting algal photosynthesis.

Drawbacks of the B.O.D. Test   The B.O.D. test as a measure of potential substrate degradation and oxygen utilization has a number of pitfalls:

1. It ignores the contribution of recalcitrant organic compounds.

2. It assumes first-order kinetics. The reaction is much more complex and does not follow first-order kinetics (Fig. 8.5). The use of L, the first-stage B.O.D., neglects the complex biological events occurring at later stages in the process.

3. It ignores the mineralization of the organic substrate and the potential for algal photosynthesis and hence further B.O.D. at a later time or place.

Uses of B.O.D. Test   The B.O.D. test is useful as a measure of potential oxygen depletion during substrate utilization in a limited period of time without taking into account the effects of released nutrients. B.O.D. provides an excellent yardstick of domestic and industrial effluents. The B.O.D. of different effluents can be compared. Effluents from the same plant can be monitored routinely. Standards are set on the basis of B.O.D. loads entering receiving waters.

STREAM PURIFICATION

Self-purification is defined as the ability of a natural body of water to rid itself of pollutants. The degradation of organic wastes is of the greatest importance. If the water is incapable of degrading the organic matter because the concentration is too high for the available oxygen, then the water becomes *anoxic*, or oxygen depleted.

In rivers the organic substrate enters the stream at a single point and is degraded in a length of the river, or *river reach*. The released nutrients do not accumulate since they are carried downstream. There is no flow of information back up the stream.

The situation in lakes and ponds is quite different. Released nutrients accumulate and algal productivity is superimposed on heterotrophic organic matter decomposition.

The Microflora The aerobic decomposition of organic matter in streams is dependent on the complex native microflora that is capable of degrading this mixture of ever-changing substrates. Two groups of microorganisms are responsible for the degradation:

1. microorganisms suspended on colloidal particles,
2. sessile microorganisms.

The vast majority of microorganisms in natural waters are not freely suspended. They are attached to detritus, clay, and other colloids in the water. This portion of the microflora lives on benthic algae and on the sediment. This group includes *Sphaerotilus*, a filamentous aerobic bacterium that attaches to surfaces in rivers containing high concentrations of organic matter. Streamers are formed from chains of cells covered in a sheath.

A photograph of *Sphaerotilus* is shown in Fig. 8.7. *Sphaerotilus* also forms large mats on the water surface that are capable of utilizing very large quantities of organic matter.

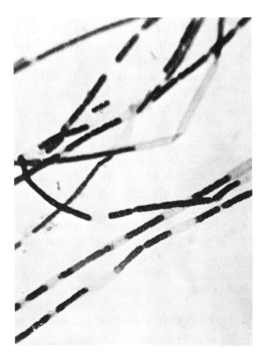

FIGURE 8.7 Photomicrograph of filaments of *Sphaerotilus natans.* Magnification 1500X. (From A. H. Romano and J. C. Peloquin, *J. Bacteriol.,* **86**:252 (1963).)

Kinetics   The chemostat is often used as a model to study organic matter decomposition in aquatic ecosystems. The kinetics of substrate utilization in the chemostat closely resemble those of streams. The substrate flows into the reactor at a known rate. Unused substrate, products, and a portion of the microflora flow out of the reactor or out of the river reach. The reaction is dependent on

1. substrate concentration,
2. flow rate or dilution rate,
3. bacterial concentration.

It can be described by the equation

$$S = Ks \left(\frac{D}{\mu m - D} \right)$$

S = substrate concentration

D = dilution rate

μm = maximum growth rate

Ks = saturation constant, equal to the substrate concentration at which $\mu = \frac{1}{2}\mu m$

The chemostat is not an ideal system for the study of stream purification.

1. It rapidly selects for a single microorganism. In contrast, rivers contain a mixed microbial population.
2. It does not support a sessile population or microorganisms growing on colloidal particles.
3. It is fully mixed. No ecological niches containing unusual microorganisms are allowed to develop. Homogeneity is the rule.
4. It cannot be used to study changing substrates. It selects for the organism that most effectively utilized the substrate flowing through the reactor. All other microorganisms are washed out. A new substrate requiring a different organism is not degraded. Natural habitats contain the diverse microflora necessary to degrade any biologically available substrate.

Mixed Substrates   It is difficult to estimate the rate of degradation of mixtures of organic substrates by heterogeneous microflora in a stream. The rate of degradation of one substrate may be affected by the presence of another substrate.

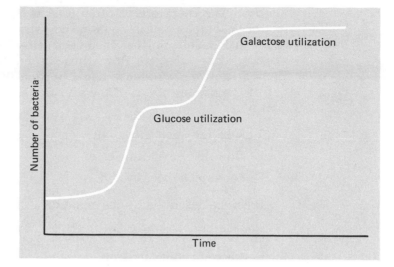

FIGURE 8.8   Catabolite repression. The production of glucosidase for glucose metabolism inhibits galactosidase. Only when the glucose has all been utilized is galactosidase produced and galactose metabolized. There is a lag period before the production of each enzyme.

Catabolite repression refers to the inhibition of certain catabolic pathways.   The metabolism of an easily utilized substrate may repress the enzymatic system necessary for utilization of a less easily metabolized substrate.   Only when the first substrate has been completely metabolized can metabolism of the second begin. Catabolite repression of galactose utilization by *E. coli* occurs in the presence of a more readily utilized substrate, glucose (Fig. 8.8). The bacterium utilizes the glucose first.   After 5 hours all the glucose has disappeared and there is a lag period before galactose metabolism begins.   The lag enables the bacterium to resynthesize the repressed enzyme β-galactosidase. These sequences of catabolite repression curves are called *diauxic growth curves.*

End product repression is an important regulatory process in the control of organic matter decomposition.   The presence of a product of metabolism frequently inhibits the synthesis of enzymes for that process (Fig. 8.9).   A negative feedback process operates in end product repression.   Examples of end product repression include the inhibition of L-histidine synthesis by *E. coli* in the presence of histidine.

An influx of L-histidine by excretion of other organisms would serve the same purpose of inhibiting histidine synthesis by *E. coli.*

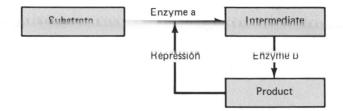

FIGURE 8.9 End product repression. Accumulation of an inter-mediate metabolite or a product inhibits the production of an enzyme in the reaction sequence.

The cell is spared the need to synthesis metabolites when they are already present.

Both catabolite and end product repression are of the utmost importance in the consideration of organic matter decomposition in natural waters. The presence of a rapidly metabolizable sub-strate may result in the accumulation by catabolic repression of a second substrate. End product repression assumes importance in situations where the products accumulate. Insufficient dilution of products can inhibit further degradation of the substrate.

Result of Purification   The end products of organic matter depletion in a stream are

1. reestablishment of autotrophic producers and a decline in heterotrophic activity,
2. reaeration of the river and reestablishment of complex communities.

The introduction of large quantities of organic matter into a stream forces the biological equilibrium in the direction of heterotrophy. The microflora becomes dominated by consumers. An indication that purification is nearing completion and that available organic substrates are being depleted is the return of photosynthetic autotrophs. Figure 8.10 shows the sequence of microorganisms along a reach of river polluted by organic matter. The initial strong heterotrophic activity is replaced downstream by photoautotrophs.

This relationship between photoautotrophs and heterotrophs has been expressed as a P/H index by Wuhrmann. He has shown the effect of different sources of organic matter on the P/H index

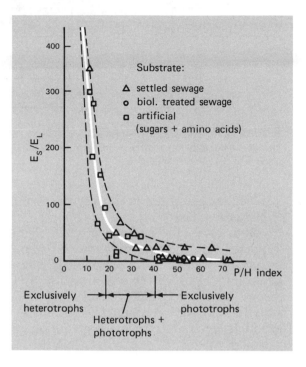

FIGURE 8.10　The effect of different sources of organic matter on the P/H index of a model river. E_S is the free energy of the organic pollutants and E_L is the influx of light energy. Both are expressed in kilocalories per square decimeter per day. There is a direct relationship between the ratio of the two energy forces entering the river and the dominance of the physiological group in the river. (From K. Wuhrmann in *Water Pollution Microbiology,* ed. R. Mitchell. John Wiley & Sons, Inc., N. Y., 1972.)

(Fig. 8.10).　The index was found to be proportional to the concentration of organic substrate and quite independent of the form or origin of the substrate.　The P/H index might prove to be a useful tool in the study of stream purification.　A stream that is free of excessive organic matter or nutrients has a moderate algal population and minimal heterotrophic microbial activity.　Unfortunately, in heavily loaded streams the inorganic nutrients released during organic matter decomposition support large growths of algae that accumulate in slow reaches of the river further downstream.　The algae act as organic substrates in these sluggish regions, causing oxygen depletion in the water and massive growths of water weeds on the water surface.

**ANAEROBIC
DECOMPOSITION**

The absence of oxygen does not inhibit microbial decomposition of organic matter. Under aerobic conditions, oxygen is the ultimate electron donor for energy transfer in the microbial cell. Under anaerobic conditions, organic compounds serve as electron donors. The energy yield in the anaerobic process is lower. This results in

1. much lower cell yields,
2. a much slower rate of decomposition,
3. incomplete degradation of the substrate.

The addition of large quantities of organic matter to a natural water results in a high rate of metabolism by aerobic microorganisms and ultimately in oxygen deficiency. Aerobic metabolism yields carbon dioxide and water. The shift to anaerobic metabolism yields organic acids. The most common are lactic, butyric, and acetic acids. A broad range of anaerobic heterotrophic microorganisms are responsible for these transformations.

Methane Formation   The formation of the flammable gas methane is one of the most important anaerobic transformations. It requires the complete absence of oxygen. These conditions are not usually met in oxygen-deficient natural waters. Methane formation is common in swamps and deep sediments. Pockets of the gas displaced into filter funnels will frequently give 12-in. flames when ignited. Rumen bacteria in cattle produce methane and are frequently used in the study of methane production.

Sludge digestion is discussed in Chapter 15. This process makes use of methane bacteria to decompose the bacterial cells produced in the secondary treatment of sewage. The anaerobic digester contains a culture of methane-forming bacteria that anaerobically degrades the sludge to form methane. The digester is maintained completely free of oxygen. The methane produced in this process is frequently sufficient to provide power for the sewage treatment plant.

The production of methane is a two-step process requiring two microorganisms (Fig. 8.11). In the first stage an organic acid or alcohol is produced anaerobically from complex organic substrates. This is not a specific process. It can be mediated by many different anaerobic microorganisms. The second step is the specific reduction of the acid or alcohol to methane by a methane bacterium.

The formation of acetic acid is an essential step in the formation of methane. One portion of the acetic acid is converted

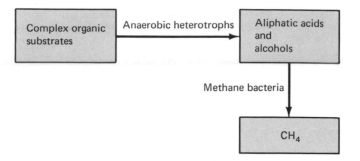

FIGURE 8.11 The formation of methane from complex organic substrates. The reaction is a two-stage anaerobic process. The first stage is nonspecific. The second stage is mediated by methane bacteria.

directly to methane and another to carbon dioxide. The carbon dioxide reacts with hydrogen produced in the formation of acetic acid to yield more methane. The reaction of ethyl alcohol is typical:

(1) $2\,CH_3CH_2OH + 2\,H_2O \rightarrow 2\,CH_3COOH + 8\,H^+$

(2) $2\,CH_3COOH \rightarrow 2\,CH_4 + 2\,CO_2$

(3) $CO_2 + 8\,H^+ \rightarrow CH_4 + 2\,H_2O$

In reaction 1, the acetic acid and hydrogen ions are formed. Reaction 2 converts acetic acid to methane and carbon dioxide. Finally, in reaction 3, the carbon dioxide and hydrogen ions react to form more methane. Each two molecules of ethanol produces three molecules of methane.

The methane-forming bacteria belong to a separate family, the Methanobacteriaceae. They are Gram-negative and are strict anaerobes. Four genera are recognized:

Methanobacterium—a nonspore-forming rod

Methanobacillus—a spore-forming rod,

Methanococcus—a nonspore-forming coccus,

Methanosarcina—a nonspore-forming coccus in packets of eight.

Figure 8.12 shows photomicrographs of some typical methane-producing bacteria. These bacteria are quite substrate specific (Table 8.3). They have been differentiated into species on the basis of the substrate used; e.g., *Methanococcus vaniellii* utilizes formate, while *M. mazeii* uses acetate or butyrate.

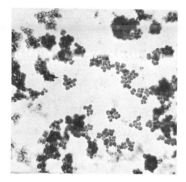

(a) Acetic acid digester.

(b) Acetic acid digester.

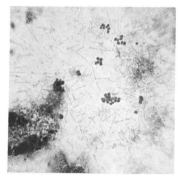

(c) Octanoic acid digester.

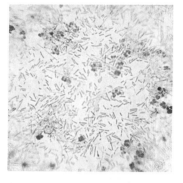

(d) Ethyl alcohol digester.

FIGURE 8.12 Photomicrographs of some methane-producing bacteria. Magnification 1000X. (Courtesy P. McCarty.)

TABLE 8.3 Specificity of methane-producing bacteria. The most common substrates are aliphatic acids and lower alcohols.

Methane-Producing Bacterium	Substrate
Rod-shaped bacteria	
Methanobacterium propopionicum	Propionate
Methanobacterium sonhgenii	Acetate, butyrate
Methanobacterium omelianskii	Primary alcohols, secondary alcohols
Coccoid bacteria in chains	
Methanococcus vaniellii	Formate
Methanococcus mazeii	Acetate, butyrate
Bacteria in packets	
Methanosarcina barkerii	Acetate, methanol

SUMMARY

1. Microorganisms utilize carbon compounds originating in dead organisms, excretions from both plants and animals, and non-biogenic organic compounds (usually hydrocarbon derivatives). Biogenic carbon is formed by the fixation of carbon dioxide by autotrophic microorganisms.

2. Decomposition of organic carbon compounds is dependent on the presence of a sufficient concentration of other elements for microbial growth. Nitrogen and phosphorus frequently are limiting for microbial growth. An adequate ratio of carbon: nitrogen:phosphorus is 100:10:1.

3. The rate of aerobic decomposition of organic matter in water is frequently estimated by measurement of the biochemical oxygen demand (B.O.D.). In running streams the B.O.D. is balanced by reaeration. This relationship is described by the Streeter-Phelps equation. Organic matter decomposition in streams is a complex ecological phenomenon. The ecological changes can be described in terms of a photoautotroph/heterotroph (P/H) index.

4. Anaerobic decomposition of organic matter is much slower than aerobic degradation. There is a lower cell yield and degradation is incomplete. Under completely anaerobic conditions, methane is formed by a specific population of methane bacteria.

FURTHER READING

M. Alexander, *Soil Microbiology*. John Wiley & Sons, Inc., New York, N.Y., 1961. See section on the carbon cycle.

A. F. Gaudy, "Biochemical Oxygen Demand" in *Water Pollution Microbiology*, ed. R. Mitchell. John Wiley & Sons, Inc., New York, N.Y., 1972.

E. Stumm-Zollinger and R. H. Harris in *Organic Compounds in Aquatic Environments*, ed. S. J. Faust and J. V. Hunter. Marcel Dekker, Inc., New York, N.Y. 1971.

C. J. Velz, *Applied Stream Purification*. John Wiley & Sons, Inc. New York, N.Y., 1070.

K. Wuhrmann, "Stream Purification" in *Water Pollution Microbiology*, ed. R. Mitchell. John Wiley & Sons, Inc., New York, N.Y., 1972.

RECALCITRANT

ORGANIC

COMPOUNDS

9

In the last chapter we discussed the microbial degradation of organic materials that are easily and rapidly decomposed. In this chapter we shall consider the man-made organic compounds that are finding their way in increasing quantities into soil and water. Many of these are strongly resistant to biodegradation, or *recalcitrant*.

The dramatic increase in technological development in the United States during the past quarter century has put an ever-increasing strain on the waters that receive recalcitrant organic compounds. These include complex organic chemicals produced as by-products in the synthetic chemical industry, in the plastics industry, together with petroleum hydrocarbons.

The agricultural industry utilizes many millions of tons of recalcitrant pesticides. These chemicals are partially responsible for our huge crop yields. Paradoxically, some of the most effective pesticides are also responsible for a health hazard to humans and a serious danger to the biosphere.

Detergents took the place of soaps as domestic cleaning agents almost 20 years ago. Recalcitrant *hard* detergents have been replaced in many countries by easily biodegraded *soft* detergents.

INDUSTRIAL WASTES

Immense quantities of organic compounds enter our natural waters in industrial effluents. The dyestuff industry yields effluents containing phenolic compounds (Table 9.1). A group of pyridenes are formed in the manufacture of coal gas from coke. Oil refineries produce benzpyrenes as waste products, and phenolic compounds are products of the paper industry. All of these organic materials are strongly recalcitrant and accumulate in natural waters. They

TABLE 9.1 Some common recalcitrant chemicals from industrial sources entering natural waters.

Source	Chemical
Plastics industry	Polychlorinated biphenyls
Dyestuff industry	Benzidenes Naphthylamines
Sugar refining	Methylamine Ethanolamine Amylamine
Coal wastes	Pyridines Anilines

are representative of an ever-widening group of industrial pollutants.

Recalcitrant organic waste products must be prevented from entering our ecosystem if ecological damage is to be prevented. A new branch of engineering technology is devoted to recovery of these organic compounds from industrial wastewaters. There is little doubt that this new technology will result in the creation of new uses for materials that were once considered to be useless waste products.

PETROLEUM HYDROCARBONS

Sources   The worldwide increase in use of petroleum products has been dramatic during the past 25 years. Much of this oil travels by ship from the oil field to its destination. In addition, increasing use is being made of overland pipes. Petroleum hydrocarbons find their way to our natural waters from the following sources:

1. shipping accidents,
2. offshore oil drilling,
3. cleaning of ships' tanks and bilges,
4. disposal of waste oils from industry and automobiles,
5. pipeline damage.

The destruction of the oil tanker Torrey Canyon in 1967 with the loss of 119,000 barrels of crude oil serves to illustrate the potential for oil loss in maritime disasters. These accidents are not confined to the sea and frequently occur in lakes and rivers. The danger of loss from offshore drilling operations was underscored in the Santa Barbara oil spill of 1969. In that incident, thousands of gallons of crude oil escaped from an offshore oil well. For more than a year, oil seeped out of the well and caused gross contamination of mainland beaches and offshore sediments.

By far the largest chronic source of oil to our natural waters comes from industry and automobiles. The quantity of oil in industrial effluents is unknown but must be significant. In the United States, 350 million gallons of automobile oil is used annually and much of the waste oil is disposed into our natural waters. Breaks in pipelines and spillage from pipeline terminals also account for a significant loss of oil. It has been estimated that there are 200,000 miles of oil pipelines in the United States carrying 1 billion tons of oil per year. Even minor leakage would result in large quantities of oil in our lakes and rivers.

A survey of organic pollutants in the Charles River in Boston in 1972 showed the presence of a mixture of recalcitrant compounds. It included alkanes, alkyl naphthalenes, alkyl anthracenes, and pyrene. The source of many of these materials was automobile exhaust condensate in runoff water.

Biodegradation Processes   Crude oil is a mixture of many different aromatic and aliphatic hydrocarbons. Each batch of oil is quite specific in its hydrocarbon composition. When a crude oil comes into contact with a natural water, the straight chain alkanes are easily degraded and most have disappeared within 7 days. Branched alkanes are more resistant and may remain for a number of months. The aromatic hydrocarbons frequently mix with the sediments before biodegradation can occur. Since hydrocarbon degradation is almost exclusively an aerobic process, the accumulation in sediments drastically lowers the rate of decomposition.

The biodegradation of hydrocarbons in natural habitats follows the following empirical rules:

1. Alkanes are more rapidly degraded than aromatic hydrocarbons.

2. Within the alkanes, straight chains are more susceptible than branched chains.

3. Chains of lengths between 10 and 18 are most rapidly oxidized. Methane, ethane, and propane are only attacked by highly specialized organisms. Waxes containing more than 30 carbons are quite insoluble and, therefore, recalcitrant.

4. Within the aromatic series, both alkyl-substituted benzenes and polycyclic compounds are more readily degraded than benzene.

The bacteria predominate in hydrocarbon degradation in natural waters. *Pseudomonas, Micrococcus, Corynebacterium,* and *Mycobacterium* are commonly involved in the oxidation process. The scanning electron micrograph in Fig. 9.1 shows a film of hexadecane completely covered by cells of the coccoid bacterium *Micrococcus certificans.* Degradation occurs even on tar balls, although at a much slower rate. Figure 9.2 illustrates that tar balls are covered with a diverse microbial population.

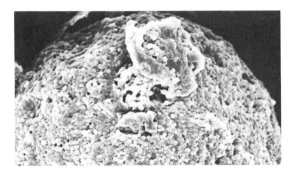

FIGURE 9.1 A scanning electron micrograph of a film of the aliphatic hydrocarbon hexadecane completely covered by the coccoid hydrocarbon-decomposing bacterium *Micrococcus certificans.* (Courtesy W. R. Finnerty.)

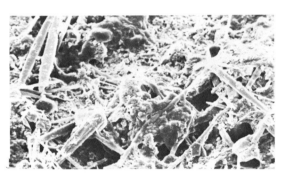

FIGURE 9.2 A scanning electron micrograph of microorganisms growing on a tar ball. (Courtesy W. R. Finnerty.)

FIGURE 9.3 Decomposition of petroleum hydrocarbons by yeasts. The kerosine globules are coated by yeast cells. (From D. Ahearn, S. P. Meyers, and P. C. Standard, *Devel. Indust. Microbiol.,* 12 (1971).)

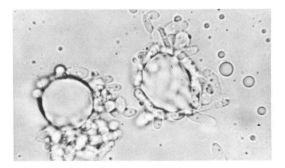

FIGURE 9.4 Emulsification of crude oil by yeasts. *Left:* The growth of the yeast on the petroleum emulsifies the oil and water mixture. *Right:* A tube containing crude oil and water remains separated in the absence of microbial growth. (Courtesy D. Ahearn.)

Yeasts are also active in the degradation of petroleum hydrocarbons. Figure 9.3 is a photograph of a hydrocarbon-decomposing yeast growing on kerosine. Emulsification of oils occurs rapidly in the presence of yeasts (Fig. 9.4), indicating the rapidity of degradation.

An adequate supply of nitrogen and phosphorus is essential for hydrocarbon degradation. These elements are present in very low concentrations in the deep ocean. Degradation of oil spilled at sea is limited by nitrogen and phosphorus deficiency. Decomposition is also strongly dependent on water temperature. At temperatures below 5°C very little degradation occurs. The optimal temperature for hydrocarbon decomposition is approximately 25°C.

Control Biological control is frequently mentioned as a means of taking care of oil spills. Hydrocarbon-decomposing microorganisms have been used in attempts to increase the rate of decomposition. This method has been only partially successful because of the inability of nonindigenous microorganisms to compete with the native microflora. Even native organisms, when added at high concentration, rapidly die back to the concentration present in the ecosystem. The development of a high population density of indigenous hydrocarbon decomposers is frequently limited by a deficiency of nitrogen and phosphorus. Addition of these elements to oil spills serves to stimulate the indigenous hydrocarbon-decomposing microflora and increase the rate of degradation.

PESTICIDES

Pesticides are used to destroy those organisms that are considered to be a threat to human life, crops, or livestock. They include insecticides, fungicides, bactericides, nematicides, and rodenticides. Use of pesticides is increasing at a rapid rate. In 1967, a billion lbs of pesticides were produced for crop protection in the United States compared to 870 million lbs in 1965. More than 300 organic pesticides are being used.

Among the fungicides, copper sulfate, pentachlorophenol, and trichlorophenol are most widely used. Common herbicides include 2,4-D, 2,4,5-T, and sodium chlorate. The aldrin group, parathions, and methyl bromide are among the most commonly applied insecticides today. DDT was used far more than any other insecticide. It is being withdrawn, however, because of the serious ecological and health hazards.

Biodegradation The structure of a pesticide determines its rate of biodegradation by the indigenous microflora. A comparison of the rate of biodegradation of two herbicides, 2,4-dichlorophenoxyacetic acid (2,4-D) and 2,4,5-trichlorophenoxyacetic acid (2,4,5-T), in soil or water is shown in Fig. 9.5. The addition of another chlorine atom in the 5 position changes the degradation time from 14 to 200 days. 2,4,5-T is quite recalcitrant to the heterotrophic population of soil and water.

FIGURE 9.5 A comparison of the rate of microbial degradation of 2,4-D and 2,4, 5-T in natural water. The addition of another chlorine to the benzene ring strongly increases the recalcitrance of the pesticide. (From M. Alexander, *Soil Microbiology.* John Wiley & Sons, Inc., N. Y., 1961. Reproduced by permission.)

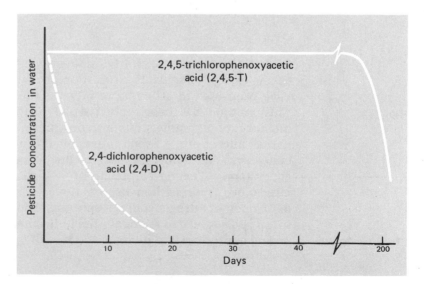

TABLE 9.2 The effect of chain length on biodegradation rates of some phenoxyalkyl carboxylic acid herbicides in soil.

Resistance to biodegradation decreases as the length of the aliphatic side chain increases.[a]

Herbicide	Aliphatic Side Chain	Days for Total Degradation in Soil
2-Chlorophenoxacetate	Acetate (C2)	205+
2-(4-Chlorophenoxypropionate)	Propionate (C3)	205+
2-(4-Chlorophenoxyvalerate)	Valerate (C5)	81+
2-(4-Chlorophenoxycaproate)	Caproate (C6)	11

[a]Reprinted from *J. Agr. Food Chem.*, 9:44 (1961). Copyright by the American Chemical Society.

TABLE 9.3 Decomposition of disubstituted benzenes by the soil microflora.

The position of chemical substituents on the benzene determines the resistance of the compound to biodegradation.[a]

Second Substituent		First Substituent							
Type	Position	COOH	OH	NO_2	NH_2	OCH_3	SO_3H	Cl	CH_3
COOH	o	2	2	8	2	4	>64	>64	16
	m	8	2	>64	>64	16	>64	32	2
	p	2	1	4	8	2	>64	64	8
OH	o		1	>64	4	4		>64	1
	m		8	4	>64	16		>64	1
	p			16		8	32	16	1
NO_2	o			>64	>64	>64	>64	>64	>64
	m			>64	>64	>64	>64	>64	>64
	p			>64	>64	>64	>64	>64	>64
NH_2	o				>64	>64	>64	>64	64
	m				>64	>64	>64	>64	8
	p					64	>64	>64	4
OCH_3	o					8			
	m					>32			
	p					8			
SO_3H	m						>64		
	p							16	24

[a]Reprinted from *J. Agr. Food Chem.*, 14:410 (1966). Copyright by the American Chemical Society.

Another example of the importance of molecular structure in recalcitrance of pesticides can be seen in a comparison of the rates of biodegradation of phenoxy pesticides with aliphatic side chains. Table 9.2 shows that the pesticide becomes more susceptible to biodegradation as the chain length of the aliphatic side chain is increased.

The position of the chemical groups on the benzene is of great importance to the degradation process (Table 9.3). Carboxyl and

FIGURE 9.6   The metabolism of DDT by microorganisms. The metabolites DDE, TDE, TDEE, and DDA are all quite recalcitrant and accumulate in higher organisms. (After F. Matsumura and G. M. Bousch in *Soil Biochemistry,* by A. D. McLaren and J. Skujins, Vol. 2, Marcel Dekker, Inc., N. Y., 1972.)

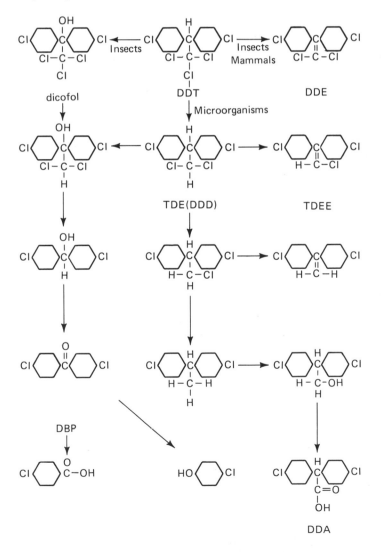

phenyl disubstituted benzenes are quite easily biodegraded. By contrast, the monochlorobenzenes and dichlorobenzenes strongly resist degradation. Pentachloronitrobenzene (PCNB) is a soil fungicide. It could be predicted that it would be extremely resistant to biodegradation. In fact no microorganisms capable of degrading PCNB have been isolated.

The threat to the environment posed by DDT and other chlorinated hydrocarbons comes from their strong resistance to microbial biodegradation. Yet, even DDT is amenable to decomposition. Figure 9.6 explains the metabolism of DDT by animals and by microorganisms. DDT is converted to TDE and TDEE in the guts of mammals by the indigenous microflora. Small quantities of DDE are also produced. The conversion to TDE is also catalyzed by soil and aquatic microorganisms. The predominant DDT decomposers are fungi. Degradation is extremely slow, in the order of years.

Recalcitrance and Structure The recalcitrance of a pesticide can be predicted on the basis of molecular structure. An approximation of comparable resistance to biodegradation as a function of structure is shown below with the least resistant materials at the top:

aliphatic acids,

organophosphates,

long-chain phenoxyaliphatic acids,

short-chain phenoxyaliphatic acids,

monosubstituted phenoxyaliphatic acids,

disubstituted phenoxyaliphatic acids,

trisubstituted phenoxyaliphatic acids,

dinitrobenzene,

chlorinated hydrocarbons (DDT).

The persistence of a number of different common pesticides in soil is seen in Table 9.4. The chlorinated hydrocarbons, represented by DDT, are by far the most resistant pesticides.

The insecticides DDT and dieldrin have residence times in soil or water of more than 3 years. The herbicides dichloropropionic acid (Dalapon) and trichloroacetic acid are biodegraded in 2 to 10 weeks. The highly toxic organophosphorus insecticides are rapidly decomposed. Parathion remains active for 20 days and Malathion only for 8 days.

TABLE 9.4   The persistence of a pesticide in soil or water is dependent on its structure.

The chlorinated hydrocarbons are extremely recalcitrant. The highly toxic organophosphorus insecticides are easily biodegraded.

Pesticide	Degradation Time in Soil or Water
Chlordane—chlorinated hydrocarbon	11 years
DDT—chlorinated hydrocarbon	3 years
Dieldrin—chlorinated hydrocarbon	3 years
Dalapon—dichloropropionic acid	10 weeks
Parathion—organophosphorus	3 weeks
Malathion—organophosphorus	1 week

There are no specific pesticide-decomposing microorganisms. The ability to degrade these materials is solely dependent on the capacity of the microorganism to synthesize the appropriate enzymes. Frequently degradation is dependent on two or more microorganisms acting sequentially. Typical pesticide-decomposing bacteria include *Pseudomonas*, *Bacillus*, *Flavobacterium*, and *Achromobacter*. *Nocardia* is a common pesticide-decomposing actinomycete and *Aspergillus* is a common fungus active against pesticides.

Environmental Conditions   Pesticide degradation, as one would expect, is strongly controlled by environmental conditions. Three parameters are of importance:

1. temperature,
2. soil type,
3. cultivation.

The effect of temperature can be seen from the observation that 92% of a mixture of aldrin and dieldrin can be detected in soil held at 7°C for 4 weeks after addition. Only 40% can be detected in soil held at 46°C.

The concentration of organic matter in the soil controls the rate of pesticide inactivation. Soils that are rich in organic matter absorb large quantities of pesticides and prevent leaching to surface waters. In contrast, sandy soils have little capacity to retain the pesticides and they rapidly find their way to surface waters.

There is a strong effect of soil cultivation on pesticide degradation. They are degraded much more rapidly in cultivated than in fallow soil. The increased rate of degradation presumably is associated with the much higher rate of microbial activity in plant root zones than in the soil.

**BIOLOGICAL
CONTROL OF PESTS**

The recalcitrance of chemical pesticides can lead to severe ecological damage. Concentration of toxic chemicals has become a severe problem that can lead to sterility of species far removed in size and location from the intended pest. In addition, pesticides that utilize their toxicity for control act on a far broader spectrum of organisms than their intended victim. A typical insecticide will kill not only the specific insect pest on a plant but many harmless other species.

Biological control depends on the ability of other organisms to eradicate the pest. Biological *second-generation pesticides* have been developed which utilize natural enemies to control specific insects and weeds which cause plant disease.

Insects can be controlled by dispersing bacteria or viruses that spread through the insect colony and destroy it. In the infection of alfalfa caterpillers by *Bacillus thuringiensis*, toxins of different strains of the bacterium have been successfully propagated to control infestations by moths. *Bacillus thuringiensis* is only pathogenic for lepidoptera. No other animals or plants are susceptible. Viruses are even more effective agents than bacteria for the control of insects. A large number of viruses against insects have been purified. These are highly specific and can be used to control an individual pest species.

Weed control can be achieved by introduction of an appropriate insect disease. The prickly pear, a cactus, covered huge areas of Australia in the early part of the century. A moth, *Cactoblastis*, which infects prickly pear, was introduced in 1925. The larvae grew so well that eggs were distributed naturally throughout Australia. The *Cactoblastis* population declined with the demise of the cactus and does not pose a threat to any other plant. Many insect infestations can be controlled by inoculation of nematodes. The DD-136 nematode shown in Fig. 9.7 infects and kills larvae, pupae, and adults of many different insects.

Soil-borne fungal diseases can be controlled by stimulating a native population of microorganisms that prey on the fungus. The fungus *Fusarium* is pathogenic on many different plants. The fungal cell wall contains chitin. This fungus can be controlled by the addition of chitin to soil in the form of lobster shells. The

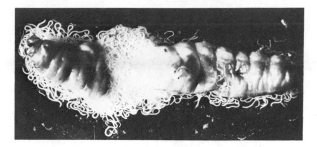

FIGURE 9.7   Infection of an insect larva with the DD-136 nematode *Neoaplectana dutkyi,* a potent weapon for the control of insect pests. (Courtesy S. Dutky.)

chitin stimulates a chitinase-producing microflora. This population degrades the chitin in the fungus cell wall, destroying the fungus.

Juvenile Hormones   *Third-generation insecticides* block one stage in the insect pest's life cycle.   Larvae produce a *juvenile hormone* that is essential for metamorphosis to the adult stage.   This hormone must be absent from the insect eggs, however, if they are to develop. If the eggs or insects come into contact with juvenile hormones, development is inhibited.   Eggs treated with juvenile hormones fail

FIGURE 9.8   Third-generation pesticides.   Juvenile hormone is required for insect larvae to develop into normal adults. When eggs come into contact with the hormone, however, a lethal derangement occurs and abnormal larvae form that fail to develop into adults.

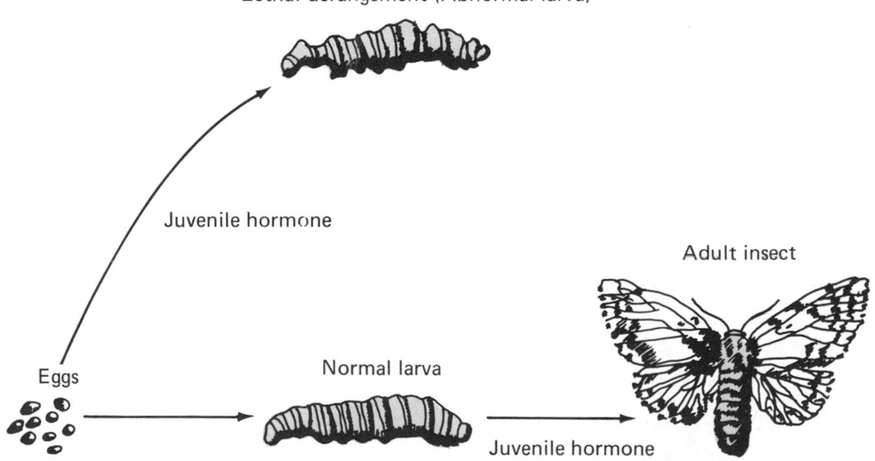

Lethal derangement (Abnormal larva)

Juvenile hormone

Adult insect

Eggs

Normal larva

Juvenile hormone

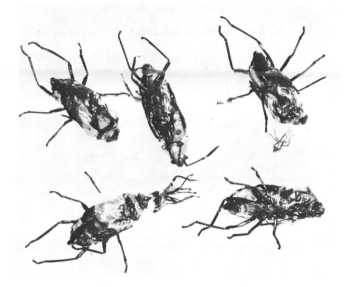

FIGURE 9.9 Effect of juvenile hormone on the metamorphosis of the final stage, milkweed bug nymphs. Abnormalities prevent development of adult animals. (Courtesy C. Williams.)

to hatch and immature insects do not undergo further development (Fig. 9.8).

Juvenile hormones have been extracted from male Cecropia moths and sprayed on eggs for biological control. Synthetic juvenile hormones have been used successfully to control Cecropia moths and the milkweed bug (Fig. 9.9). The hormones are highly species specific so that the ecological danger from their use is minimal.

BIOMAGNIFICATION

Pesticides Chemicals that are present in very low concentration in natural waters appear in high concentration in the biota. They are *biomagnified*. DDT offers a classic example of biomagnification (p. 11). Extremely low concentrations of DDT are found in natural waters throughout the world. The eggs of birds living off the fish in these waters often have sufficiently high concentrations of DDT to adversely affect reproductive ability and to endanger many species.

Petroleum Hydrocarbons are biomagnified in a similar manner. The concentration of the carcinogenic aromatic hydrocarbon 3,4-benzpyrene in contaminated shellfish can be seen in Table 9.5. The concentration in oysters in contaminated French waters

TABLE 9.5 Concentrations of 3,4-benzpyrene in contaminated shellfish.[a]

Shellfish	Location	Benzpyrene Concentration (µg/kg)
Oysters	Norfolk, Virginia	10-20
Oysters	French coast	1-70
Mussels	French coast	2-30

[a]From C. ZoBell, Proc. Joint API-EPA-USCG Conference on Prevention and Control of Oil Spills, Washington, D.C., 1971.

reached 70 µg/kg of shellfish. These levels of this strong carcinogen in human foods are hazardous.

It is commonly assumed that toxic aromatic chemicals move up the food chain through the phytoplankton and zooplankton. The photograph in Fig. 9.10 indicates that hydrocarbons are also accumulated in bacteria. The role played by microorganisms in biomagnification is still unknown.

PCB's   Polychlorinated biphenyl compounds, the so-called PCB's are appearing at an increasing rate in our natural waters. These materials, which are products of the plastics industry are extremely

FIGURE 9.10   A thin section of Micrococcus certificans grown on hexadecane. The inclusion bodies are globules of hexadecane. Magnification 165,000X. (Courtesy W. R. Finnerty and B. O. Spurlock.)

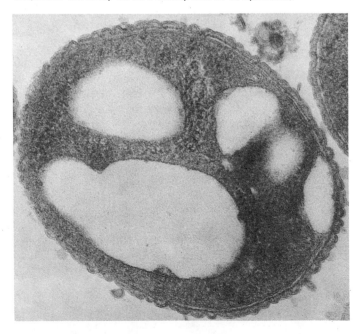

TABLE 9.6   Polychlorinated biphenyls (PCB's) in lake trout as a function of age.[a]

Age (yr)	PCB (ppm)
1	0.6-1.6
2	1.3-2.5
4	3.5-5.1
6	3.4-9.7
9	4.5-17.5
12	7.4-26.2

[a]From Bache, et al., *Science,* 177:1191 (1972). Copyright by the American Association for the Advancement of Science.

recalcitrant and are biomagnified. Table 9.6 illustrates the concentration of PCB's in lake trout as a function of age. A year-old trout may contain as little as 0.6 ppm of PCB. By the age of four, the concentration has risen to 3-5 ppm. A 12-year-old trout may concentrate as much as 26 ppm of polychlorinated biphenyl compounds. Biphenyl compounds are both human carcinogens and adversely affect the reproductive capacity of the organisms in which they are concentrated.

Direct Uptake   It is generally assumed that biomagnification of toxic chemicals occurs in a series of steps up the food chain. Concentration may also occur, however, by direct uptake from the

FIGURE 9.11   Direct uptake of polychlorinated biphenyl compounds from seawater. Intermediate steps up the food chain are not involved in this biomagnification process. (Courtesy F. Walsh.)

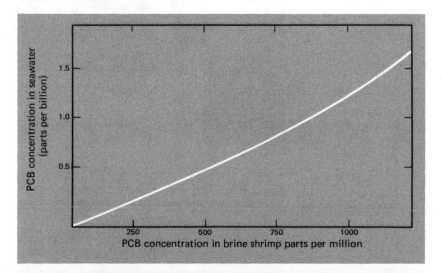

water. Most aquatic organisms pass huge quantities of water through their bodies daily. The lipids in fish gills or in filter-feeding organs are capable of absorbing directly such lipid-soluble toxins as DDT or PCB's. Figure 9.11 illustrates the direct uptake of PCB's by brine shrimp from seawater. No intermediate steps are involved and the process is quite efficient.

DETERGENTS

Soap was used as the prime cleaning agent for clothes until the introduction of detergents about 20 years ago. Detergents have much stronger surface activity and so are more efficient than soaps. Soaps are biologically *soft*. They are easily degraded by microorganisms. By contrast, the early detergents were *hard*. They retained their surface activity in the treatment facilities and in receiving waters. The result was massive bubbling, as shown in Fig. 9.12. Conversely, *soft* detergents are easily degraded and cause no aesthetic problems.

The chemical structures of hard and soft detergents are shown in Fig. 9.13. The branched alkyl benzyl sulfonates persist for more than 800 hours in natural waters. Linear alkyl benzyl sulfonates are much softer and are rapidly degraded.

FIGURE 9.12 A photograph of untreated domestic sewage containing foaming hard detergents as it enters a stream. (Courtesy Massachusetts Audubon Society.)

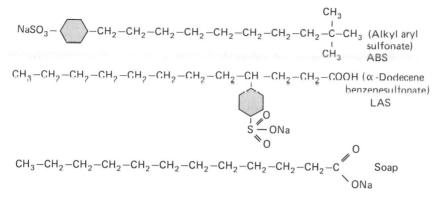

FIGURE 9.13 The chemical structure of hard and soft detergents. ABS detergents are very recalcitrant. LAS detergents and soaps are easily degraded.

Two major groups of microorganisms are responsible for detergent decomposition, *Nocardia* and *Pseudomonas*. Both are common in sewage treatment plants and in natural waters.

SUMMARY

1. Recalcitrant chemicals are those organic compounds that are not easily degraded by microorganisms. The major sources of recalcitrant materials to our environment are petroleum hydrocarbons, pesticides, and industrial effluents.

2. Petroleum hydrocarbons are a major source of marine pollution. Straight-chain alkanes are degraded more rapidly than branched alkanes. Aromatic hydrocarbons are strongly recalcitrant and frequently accumulate in the sediment.

3. The rate of degradation of pesticides is dependent on their structure. Aliphatic acids and organophosphate pesticides are easily degraded. An intermediate group includes unsubstituted and monosubstituted phenoxyalphatic acids. Disubstituted and trisubstituted phenoxyaliphatic acids remain undegraded for months. Chlorinated hydrocarbons, including DDT, are extremely recalcitrant and remain untouched by microorganisms for years. Biological pesticides are being considered as an alternative to the use of toxic chemicals for pest control.

4. The fate of recalcitrant organic compounds is often to be biomagnified. Both the pesticide DDT and the components of

plastics, polychlorinated biphenyl compounds (PCB's), are biomagnified. These and other recalcitrant organic compounds accumulate in sufficiently high concentrations to pose a threat to animals and to humans.

5. Hard detergents accumulate in natural waters, causing aesthetic problems. Soft detergents are rapidly degraded either in treatment plants or in receiving waters.

FURTHER READING

American Chemical Society, "Organic Pesticides in the Environment," *Advances in Chemistry Series*, No. 60, 1966.

"Cleaning our Environment, the Chemical Basis for Action." A report of the American Chemical Society, Washington, D.C., 1969.

P. De Bach, *Biological Control of Insects, Pests, and Weeds.* Reinhold Pub. Co., New York, N.Y., 1964.

G. D. Floodgate, "Biodegradation of Hydrocarbons in the Sea" in *Water Pollution Microbiology*, ed. R. Mitchell. John Wiley & Sons, Inc., New York, N.Y., 1972.

D. D. Kaufman and J. P. Plimmer, "Approaches to the Synthesis of Soft Pesticides" in *Water Pollution Microbiology*, ed. R. Mitchell. John Wiley & Sons, Inc., New York, N.Y., 1972.

R. L. Kolpack, "Biological and Oceanographical Survey of the Santa Barbara Channel Oil Spill, 1969-70," Hancock Foundation, University of S. California, Calif., 1971.

R. White-Stevens, *Pesticides in the Environment.* Marcel Dekker, Inc., New York, N.Y., 1971.

C. Williams, "Third Generation Pesticides" in *Man and the Ecosphere, Readings from Scientific American.* W. H. Freeman Co., San Francisco, Calif., 1971.

U.S. Dept. of the Interior, "Oil Pollution, A Report to the President," Washington, D.C., 1968.

NUTRIENT CYCLES

The regeneration of nutrients is one of the most essential functions of microorganisms in the biosphere. The primary producers discussed in Chapter 1 remove an infinitely small quantity of minerals from the huge reservoir in rocks and sediments. The vast bulk of the minerals utilized by the world's biota is recycled. Organisms die and are mineralized to release inorganic chemicals. These chemicals are utilized by other organisms in the biosphere. Odum has differentiated between the large abiotic reservoir pool of nutrients, which may be unavailable to organisms, and the exchangeable pool, which is available. It is this available pool that is recycled by the native microflora.

Energy flows from both the large abiotic and the small biotic nutrient pool (Fig. 10.1). The autotrophs, or producers, are responsible for the release of inorganic nutrients from mineral sources. Once the minerals have entered the exchangeable pool, both autotrophs and heterotrophs are involved in the recycling process.

The cycle of carbon is discussed in Chapter 8. In this chapter, I intend to explain the recycle pathways of nitrogen, phosphorus, and sulfur.

TRANSFORMATIONS OF NITROGEN

Nitrogen is present in biological systems in large quantities. It is present in a wide range of cellular materials including proteins and nucleic acids; yet nitrogen is frequently deficient in the biosphere. We must be concerned with the complex cycle of nitrogen in soil and water that is controlled by the microflora. The availability of nitrogen for crop or algal growth is governed by these processes. Hence they are of the utmost importance in controlling biological activity.

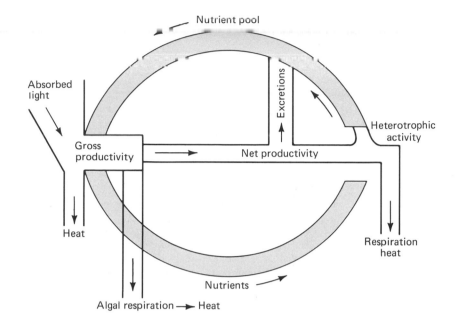

FIGURE 10.1 A biogeochemical cycle superimposed on an energy flow diagram. The materials are recycled in contrast to the flow of energy, which is unidirectional. Net productivity is the biomass of primary production less that consumed by heterotrophs. (After E. P. Odum, *Fundamentals of Ecology,* 3rd ed. W. B. Saunders Co., Philadelphia, Pa., 1971.)

Figure 10.2 illustrates the transformations of nitrogen. There are five important processes occurring in nature:

1. nitrogen immobilization,
2. nitrogen mineralization,
3. nitrification,
4. denitrification,
5. nitrogen fixation.

Biodegradation of animal, plant, and microbial cells or their organic excreta is accompanied by the conversion, or *mineralization*, of organic nitrogen to the inorganic form. Conversely, during growth of plant or microbial cells, inorganic nitrogen is assimilated by being converted into organic protoplasm.

The inorganic forms of nitrogen are interconverted in a complex series of microbial processes. Nitrification describes the conversion of ammonia to nitrate, which occurs under highly

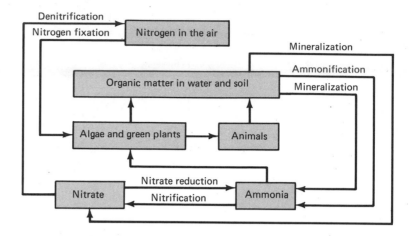

FIGURE 10.2 The nitrogen cycle. Nitrogen is absorbed from the air, soil, and water by algae and plants. It is transformed by heterotrophic micro-organisms. Five important transformations occur: mineralization, assimilation, nitrification, denitrification, and nitrogen fixation.

aerobic conditions. In the total absence of oxygen, nitrate is converted to ammonia. Denitrification of nitrate to nitrogen gas and nitrous oxide occurs under reducing but not totally anaerobic conditions. Nitrogen gas is pumped back into the system from the air by the nitrogen-fixing bacteria that are capable of utilizing the nitrogen gas in the air.

Immobilization　Microbial growth involves the utilization of carbon substrates. For each 100 units of carbon incorporated into the cell, approximately 5 to 10 units of nitrogen are incorporated. The C:N ratio of the bacterial cell is between 5 and 10. This uptake by microorganisms of mineral nitrogen in the form of ammonium or nitrate is known as *immobilization* or *assimilation*.

Nitrogen is frequently deficient in soil and natural waters. Nitrogen uptake and growth of microorganisms occurs until the C:N ratio of the environment is reduced below 10. At that point, growth ceases because of carbon deficiency. The bacteria, actinomycetes, and fungi compete successfully with the algae for nitrogen because of their more rapid growth rate. For algal growth to occur there must be sufficient nitrogen present for all the micro-flora. In the open oceans, which are nutrient deserts, both primary productivity and heterotrophic growth are limited by nitrogen and phosphorus deficiency.

Mineralization   Mineralization of nitrogen involves the release of inorganic forms of nitrogen during microbial degradation of organic nitrogenous compounds. The release of inorganic nitrogen compounds is an essential part of the recycling process. All animal, plant, and microbial cells as well as organic excretions of these organisms are ultimately degraded by the soil and aquatic microflora. The most common nitrogen compounds to be mineralized are the proteins. Their degradation releases ammonium ions.

As in the case of immobilization, the mineralization process is dependent on the concentration of carbon in the ecosystem. Release of inorganic nitrogen is proportional to release of inorganic carbon as CO_2. The C:N ratio of mineralized products is usually 5 to 15.

Nitrification   Ammonia released during mineralization of organic matter is oxidized to nitrate by nitrifying bacteria in a two-step process:

$$\underset{NH_4^+}{\underline{\text{Ammonium}}} \xrightarrow{\quad\textit{Nitrosomonas}\quad} \underset{NO_2^-}{\underline{\text{Nitrite}}} \xrightarrow{\quad\textit{Nitrobacter}\quad} \underset{NO_3^-}{\underline{\text{Nitrate}}}$$

Nitrosomonas and *Nitrobacter* are Gram-negative nonspore-forming rods that are highly aerobic. They are both obligately chemoautotrophic, differing only in the ability of *Nitrosomonas* to utilize NH_3 as an energy source, while *Nitrobacter* utilizes NO_2^-.

There are many common nitrifying bacteria other than *Nitrosomonas* and *Nitrobacter*. They are differentiated into genera on the basis of morphology and the composition of their nucleic acids. Figure 10.3 shows electron micrographs of different morphological forms of ammonia-oxidizing bacteria, and Fig. 10.4 illustrates some common nitrite-oxidizing bacteria.

The nitrifying bacteria are highly sensitive to environmental conditions.

1. They are stricly aerobic and require high levels of oxygen.

2. *Nitrobacter* is inhibited at pH values above 9.5 in the presence of NH_4^+. This leads to accumulation of toxic nitrites under alkaline conditions. *Nitrosomonas* is active under alkaline conditions but is inhibited at pH values below 6.0.

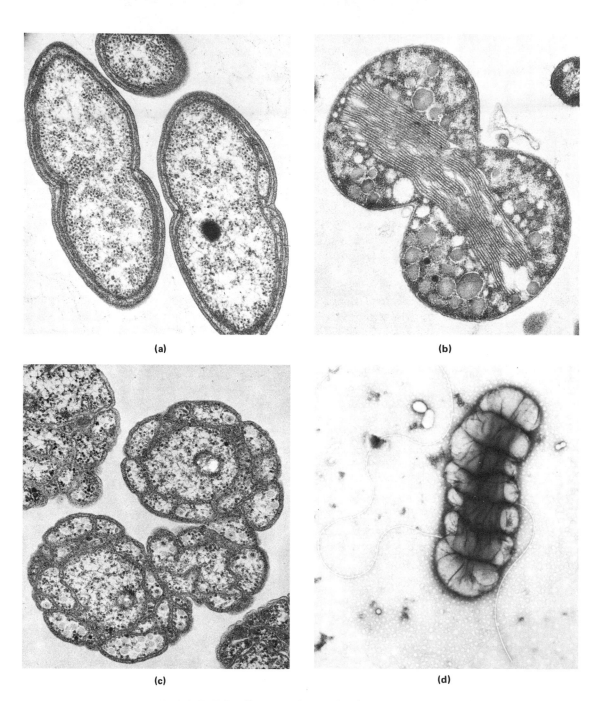

FIGURE 10.3 Electron micrographs showing different morphological forms of ammonia-oxidizing bacteria. (a) *Nitrosomonas* showing peripheral cytomembranes. (b) *Nitrocystis* showing vesicles in the cytoplasm. (c) *Nitrosolobus* forms lobular cells. (d) *Nitrospira briensis,* a spiral motile bacterium. (From S. W. Watson and M. Mandel, *J. Bacteriol.,* 107:563 (1971).)

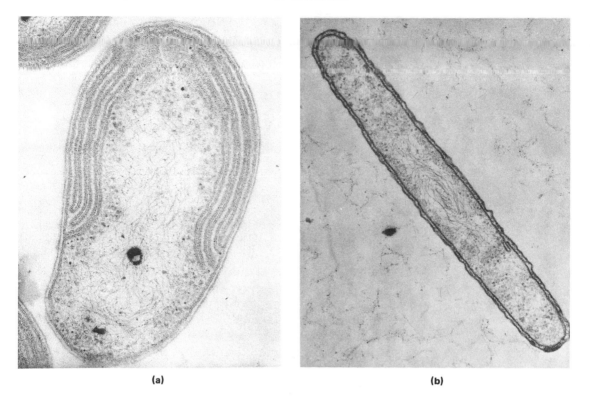

(a)   (b)

FIGURE 10.4   Electron micrographs showing different morphological forms of nitrite-oxidizing bacteria. (a) *Nitrobacter,* the best known form. (b) *Nitrospina gracilis* forms long, slender rods. (From S. W. Watson and M. Mandel, *J. Bacteriol.,* 107:563 (1971).)

3. The temperature optimum for nitrification is 30°C. No activity is detected below 5°C or above 40°C.

Nitrification occurs in the presence of organic nitrogen despite inhibition of autotrophic nitrifying bacteria. A number of heterotrophic microorganisms are capable of catalyzing nitrification. These include *Pseudomonas*, *Bacillus*, *Nocardia*, and *Streptomyces*.

Dentrification   In the absence of oxygen many heterotrophic bacteria utilize nitrate as an electron acceptor. These bacteria are known as *denitrifiers*. Denitrification occurs in waterlogged soils and in natural waters that have become deficient in oxygen.

The denitrification process involves reduction of nitrate to nitrogen gas and nitrous oxide.

Nitric acid	Nitrous acid	Nitrous oxide	Nitrogen gas

$$HNO_3 \longrightarrow HNO_2 \longrightarrow N_2O \longrightarrow N_2$$

As the system becomes more oxygen limited, a higher percentage of N_2 is produced.

The most common denitrifying bacteria are

Bacillus denitrificans,

Micrococcus denitrificans,

Pseudomonas stutzeni,

Achromobacter.

Denitrification occurs under a wide range of environmental conditions. Factors that control the process include the following:

1. *Substrate.* Denitrification by a heterotrophic microflora occurs in the presence of available organic substrates. Typical among this group are *Pseudomonas denitrificans* and *Achromobacter.* In the absence of organic carbon compounds the autotrophic denitrifiers *Thiobacillus denitrificans* and a facultative autotroph *Micrococcus denitrificans* predominate.

2. *Oxygen.* The oxygen requirements for denitrification are complex. It is necessary that the system not be totally anaerobic. Initially reduced nitrogen compounds should be oxidized to nitrate. This process requires highly aerobic conditions. The nitrate must be carried to anaerobic zones for denitrification to occur. In natural waters nitrification occurs at the surface. Denitrification occurs when the nitrate is carried to oxygen-deficient, zones in the water column, usually at the water-sediment interface.

3. *Temperature.* The optimum temperature is $25°C$. Denitrification has a wide temperature tolerance from 2 to $60°C$. Thus nitrogen loss by volatilization continues in anoxic waters even in the winter, although at a slower rate.

4. *pH.* Denitrification is inhibited under acidic conditions. At pH values of less than 5.0, no denitrification occurs. Below pH 6.0, nitrogen production is inhibited and N_2O is the only gas produced.

Nitrogen Fixation. The nitrogen-fixing bacteria absorb nitrogen gas from the air and convert it to cell protein. There are three groups of nitrogen-fixing microorganisms:

1. nonsymbiotic bacteria,
2. blue-green algae,
3. symbiotic bacteria.

The major genera of nonsymbiotic nitrogen-fixing bacteria are *Azotobacter* and *Clostridium*. *Azotobacter* is common in fertile soil. It is an aerobic, Gram-negative nonspore-forming rod. Its anaerobic counterpart *Clostridium*, which is active in swamps and sediments, is a spore-forming Gram-positive rod.)

Some of the blue-green algae, particularly members of the filamentous Nostocales family, fix nitrogen in natural waters. The two best known members of this family are *Anabaena* (Fig. 10.5) and *Nostoc*.

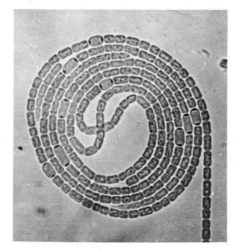

FIGURE 10.5 A nitrogen-fixing blue-green alga, *Anabaena cylindrica*. (Courtesy A. Walsby.)

Symbiotic nitrogen fixation occurs when the roots of leguminous plants become infected with the common soil bacterium *Rhizobium*. The infection results in the formation of nodules on the plant roots (Fig. 10.6a). These nodules, which are filled with *Rhizobium* cells (Fig. 10.6b), fix atmospheric nitrogen to yield organic nitrogen compounds for the bacteria and plant cells.

This process is of great importance in agriculture. Farmers grow legumes and plow in the roots as a means of maintaining the nitrogen level of the soil. Nitrogen-fixing bacteria can also utilize ammonia or nitrate. In the presence of these forms of nitrogen, however, nitrogenase, which is an essential enzyme in the conversion of N_2 to NH_3, is inhibited.

(a)

FIGURE 10.6 Symbiosis between a legume and a bacterium. (a) Nodules, about 3-4 mm in diameter on the roots of soybeans. (b) An electron micrograph of a portion of a nodule cell. The host cell is packed with envelopes containing cells of the nitrogen-fixing bacterium *Rhizobium*. (Courtesy A. Bergersen.)

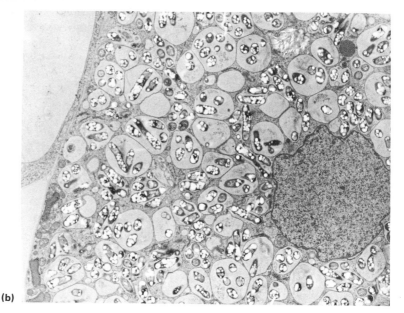

(b)

NITRATE TOXICITY

Nitrate Toxicity Nitrogen occurs in drinking water mainly as nitrate. The concentration is rarely above 1.0 ppm. In agricultural areas where nitrogen fertilizers percolate into the groundwater, however, the concentration of nitrate can reach as high as 1000 ppm. Drinking water is not a serious potential source of nitrate toxicity in adults. They can tolerate an intake of at least 0.4 mg/kg of body weight per day, but concentrations of nitrate in drinking water between 90 and 140 ppm have been implicated in the disease *methemoglobinemia* of infants under 4 months of age. The disease is characterized by the inability of the red blood cells to carry oxygen and leads to asphyxia.

Infants have a lower acidity in their stomachs than adults. This permits the development of a microbial population that reduces the nitrate to nitrite. It is the nitrite that is toxic. The toxic process is illustrated in Fig. 10.7. The respiratory pigment, hemoglobin, contains reduced iron, Fe^{2+}. This pigment is essential for the transport of oxygen in the bloodstream. The oxidized hemoglobin, oxyhemoglobin, is a red color. In the presence of nitrite the iron is oxidized to the ferric form and the hemoglobin forms a brown pigment, methemoglobin, which is incapable of transporting oxygen. When the methemoglobin level in the blood reaches 70%, asphyxia occurs.

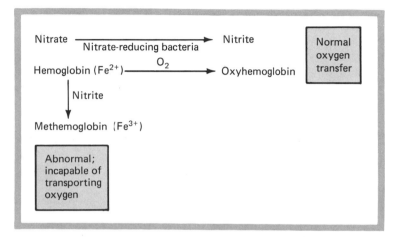

FIGURE 10.7   The production of methemoglobinemia in infants.

Nitrosamines   The group of *N*-nitroso compounds known as nitrosamines (Fig. 10.8) are highly carcinogenic. They may also be mutagenic and teratogenic. When rats are fed 5 ppm of *N*-nitrosodimethylamine, they develop tumors. At high acidity the human intestinal microflora is capable of catalyzing the reaction shown in Fig. 10.8 between nitrites and secondary and tertiary

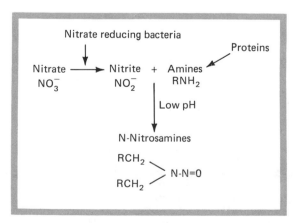

FIGURE 10.8   Formation of nitrosamines in the human intestine. Nitrates are reduced to nitrites by the intestinal microflora. At low pH values nitrites react with amines to form the carcinogenic N-Nitrosamines.

amines produced from proteins to yield nitrosamines. The processes involved are not understood; however, the potential danger of nitrates as precursors of human carcinogens suggests that they should be held to an absolute minimum in drinking waters.

The United States Public Health Services has set 45 ppm as the maximum tolerance level for nitrate in drinking water.

PHOSPHORUS

Phosphorus is an essential element in all biological systems. The most common use of phosphorus is in the energy-rich phosphate bonds of adenosine triphosphate (ATP) in the nucleic acids and the phospholipids. The ratio of carbon to phosphorus in cells is approximately 100:1. Phosphorus is frequently in short supply in soil and water so that primary productivity and sometimes even biodegradation by microorganisms is limited.

Phosphorus originates in rocks and sediments. It is mined for use in fertilizers, detergents, and many industrial processes. It is continually recycled in our biological processes. At each turn of the cycle, however, a portion of the mineralized phosphorus is rediluted in our lakes and oceans or is redeposited in the sediments of lake or ocean floors.

The phosphorus cycle is confined to those microbial transformations that mineralize organic phosphorus, solubilize insol-

FIGURE 10.9   The phosphorus cycle in natural waters. Most of the phosphorus is found in the sediments. Exchange occurs between the sediment and water column. The phosphorus is immobilized by organisms growing in the water and mineralized when the heterotrophic microflora consumes the other aquatic organisms.

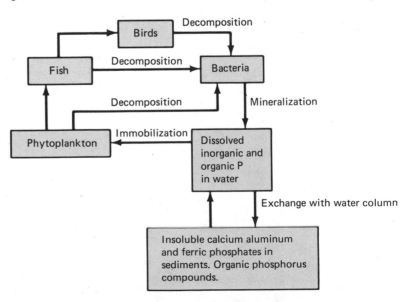

uble forms, or assimilate inorganic phosphates to cell material. The cycle is summarized in Fig. 10.9.

Mineralization   All living material contains organic phosphorus compounds. Phosphorus mineralization, resulting in release of inorganic forms, is intimately connected to microbial degradation of organic matter. It is quite nonspecific and is mediated by bacteria, actinomycetes, and fungi. The controlling factor is the ability of the microorganism to utilize the carbon substrate. The release of soluble phosphate is a by-product of this reaction.

Since phosphorus mineralization is dependent on organic matter decomposition, it is controlled by those factors that govern the degradation of organic matter. Important factors include

1. an easily degraded carbon substrate,
2. an adequate nitrogen supply,
3. appropriate pH and temperature.

A good example of phosphorus mineralization is found in a eutrophic lake in the autumn. The temperature may have fallen below the optimum for algal productivity. The indigenous microflora utilize the algae as a source of organic carbon, releasing inorganic nitrogen and phosphorus compounds. Both carbon and nitrogen supplies are adequate and the temperature and pH are appropriate for substrate degradation.

Solubilization   There is a continuous exchange between phosphorus in solution and in sediments. As the element is taken out of solution by the biota more comes into solution from the sediments.

The autotrophic bacteria living on sediments are of the greatest importance in phosphate solubilization. The nitrifying bacteria mediate the formation of nitric acid:

$$\frac{\text{Ammonium}}{NH_4{}^+} \xrightarrow{\text{nitrification}} \frac{\text{Nitric acid}}{HNO_3}$$

The sulfur bacteria mediate the production of sulfuric acid:

$$\text{Reduced sulfur compounds} \xrightarrow[\text{bacteria}]{\text{sulfur}} \frac{\text{Sulfuric acid}}{H_2SO_4}$$

The equilibrium may be described by the equation

Phosphate in solid phase $\rightleftharpoons$ phosphorus in liquid phase
(Aluminum + calcium salts)

The solubilization process is accelerated by the production of acidity on the sediment surface. Any microorganism that utilizes a substrate at the sediment-water interface and produces acidity will accelerate phosphate solubilization by producing the soluble orthophosphate ions $H_2PO_4^-$ or HPO_4^{2-}.

Assimilation Inorganic phosphates are transformed to organic phosphorus compounds primarily by assimilation into microbial cells. In the aqueous environment, algae are the prime absorbants. In soil, bacteria fix large quantities of phosphorus. Most natural environments are deficient in phosphates and the addition of available forms stimulates either bacterial or algal growth. In agricultural lands sufficient phosphorus must be added in fertilizers to provide both the microflora and the crop with an abundance of phosphate. The microflora competes successfully with plant roots for an inadequate supply.

Frequently phosphorus assimilation is inhibited by some other environmental limitation to microbial growth. When phosphate is released following algal death in ponds in the autumn, reassimilation to either bacterial or algal cells must await the return of appropriate temperature or light conditions. Assimilation requires an appropriate level or organic carbon for heterotrophs and of available inorganic nitrogen for both heterotrophs and autotrophs. When the C:P ratio declines below 100:1, then immobilization of phosphorus by heterotrophs stops and mineralization begins as cells begin to die and release inorganic phosphorus. The abundance of CO_2 in the air and dissolved in water prevents a similar restriction to algal growth. Phosphorus assimilation requires a minimal N:P ratio of 10:1. At ratios less than this, assimilation stops and both inorganic nitrogen and phosphorus are released.

SULFUR

Sulfur is found in abundance in minerals and sediments and in soil and water. It rarely limits microbial or plant growth despite its presence in all living organisms. The sulfur amino acids are essential components of cell proteins. Cysteine, cystine, and methionine are sulfur amino acids. The ratio of carbon to sulfur in microbial cells is similar to the C:P ratio of 100:1.

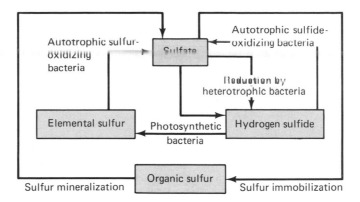

FIGURE 10.10 The sulfur cycle. Four important microbial transformations occur: mineralization of organic sulfur compounds, assimilation of inorganic sulfur, oxidation of reduced inorganic sulfur compounds, and reduction of oxidized sulfur compounds.

Sulfur transformations can be divided into four groups (Fig. 10.10):

1. mineralization of organic forms,
2. assimilation of inorganic sulfur compounds,
3. oxidation of reduced inorganic sulfur compounds,
4. reduction of oxidized inorganic sulfur compounds.

Mineralization The organic forms of sulfur are mineralized by microbial degradation of dying and dead cells or organic excretions containing sulfur compounds. Human excreta contains about 1% organic sulfur, mainly in the bacterial cells. Release of inorganic sulfur is quite nonspecific and is dependent on the microflora utilizing the carbon substrate. Sulfur release is dependent on there being an adequate supply of sulfur in either the carbon substrate or the surrounding environment for microbial growth. In practice there is usually a plentiful supply of sulfur. Mineralization is usually limited by deficiencies in available carbon, nitrogen, or phosphorus.

The most abundant inorganic forms of sulfur produced by mineralization include hydrogen sulfide, elemental sulfur, thiosulfate, and sulfur. Other inorganic sulfur compounds are produced in small quantities.

Assimilation Inorganic sulfur is assimilated by microorganisms in many different forms. All microorganisms, including algae, can utilize inorganic sulfur. Anaerobes use hydrogen sulfide or sulfur amino acids as their sulfur source. Aerobes utilize the more oxidized forms. The abundant supply of inorganic sulfur in soil and water allows organisms the choice of absorbing different forms of the element.

Oxidation Reactions   The oxidation of reduced forms of sulfur is mediated by specific sulfur bacteria. Some of these bacteria are heterotrophic, while others are chemoautotrophic, using reduced sulfur compounds as their sole energy source. Two groups of sulfur-oxidizing bacteria are recognized.

1. *Thiobacillus*,
2. filamentous bacteria.

The thiobacilli are either obligately or facultatively autotrophic bacteria. They are short Gram-negative nonspore-forming rods. Many species are differentiated on the basis of the sulfur compound they use as an energy source. The best known sulfur bacterium is *Thiobacillus thiooxidans*. It is a highly aerobic strict autotroph and converts elemental sulfur to sulfuric acid. It has the ability to grow and metabolize in highly acidic conditions. It grows well at pH 2. *Thiobacillus novellus* utilizes thiosulfate as its energy source. It is more sensitive to acidity and dies below pH 4. *Thiobacillus denitrificans* can grow anaerobically. It uses nitrate as the terminal electron acceptor and sulfur as the energy source to produce sulfates and nitrogen gas as products:

$$5\,S + 6\,KNO_3 + 2\,H_2O \rightarrow K_2SO_4 + 4\,KHSO_4 + 3\,N_2$$

The thiobacilli are active in unpolluted waters that are low in organic matter. Under these conditions autotrophic microorganisms are abundant.

The filamentous sulfur bacteria are highly aerobic chemoautotrophs. They utilize hydrogen sulfide as an energy source and deposit sulfur inclusions in the cells. They are found in large numbers as a film on the surface of deep lakes and muddy ponds where hydrogen sulfide is bubbling to the surface. They are particularly abundant in sulfur springs. Two genera are common, *Beggiotoa* and *Thiothrix*.

Reduction Reactions   In anaerobic sediments, waterlogged soils, and in waters that have become oxygen deficient because of pollution, oxidized sulfur compounds are reduced to hydrogen sulfide. This process is highly specific. It is mediated by *Desulfovibrio*, a strictly anaerobic Gram-negative nonspore-forming rod. This bacterium grows heterotrophically on many different organic substrates It can use sulfate, thiosulfate, or elemental sulfur as electron acceptors. The production of hydrogen sulfide by *Desulfovibrio* is also associated with iron reduction and the corrosion of iron pipes. This process is discussed on p. 213.

SUMMARY

1. Nutrient regeneration is an essential process in the biosphere. It is achieved by microbiological transformations.

2. Nitrogen transformations include mineralization, assimilation, nitrification, denitrification, and nitrogen fixation.

3. Nitrate toxicity occurs when water or food containing high concentrations of nitrate are consumed. Nitrate and nitrite may also be carcinogenic.

4. Phosphorus transformations by microorganisms are mainly concerned with mineralization, solubilization, and assimilation.

5. The sulfur cycle includes assimilation, mineralization, oxidation of reduced inorganic sulfur compounds, and reduction of oxidized sulfur compounds.

FURTHER READING

M. Alexander, *Soil Microbiology*, John Wiley & Sons, Inc., New York, N. Y., 1971. See chapters on nitrogen, phosphorus, and sulfur transformations.

E. Odum, *Fundamentals of Ecology*, 3rd ed. W. B. Saunders Co., Philadelphia, Pa., 1971. See chapter on biogeochemical cycles.

EUTROPHICATION

In the aquatic environment, algae are the primary producers. Growth occurs in a narrow *euphotic* zone near the surface, where the light intensity is sufficient to support photosynthesis. Productivity of algae is controlled primarily by the intensity of light and the abundance of inorganic nutrients. An excess of nitrogen or phosphorus is the most common cause of excessive algal growth. These and other factors controlling algal productivity are discussed below.

AQUATIC HABITATS

Rivers In fast-running reaches of rivers the dominant algae are the *plankton*. These are unicellular and are free floating. As the current slows down, however, a sessile population predominates. These are predominantly plants that attach to the sediment. They include the water lilies, bulrushes and the common pond weed *Potomageton*. A nutrient-rich stretch of a slow-moving stream is shown in Fig. 11.1.

Lakes In shallow lakes or ponds the euphotic zone reaches the bottom and there is a minimal temperature difference throughout the water column. The algal and plant community behaves in a similar manner to a slow-moving stream. In deep lakes, where light does not penetrate to the deep layers, an entirely different situation exists. This can be seen diagrammatically in Fig. 11.2.

During the warm summer months the sun heats the upper layers of the water causing circulation to a depth known as the *thermocline*. The zone of well-aerated warm water above the thermocline is called the *epilimnion*. The cold noncirculating

FIGURE 11.1 A nutrient-rich stretch of a slow-moving stream. Excessive productivity is in the form of water weeds. (Courtesy Massachusetts Audubon Society.)

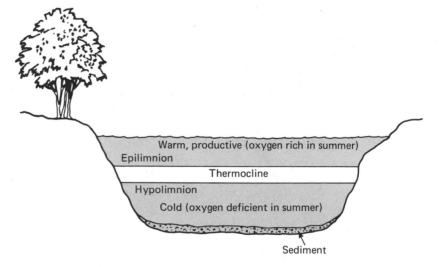

FIGURE 11.2 Thermal stratification of a deep lake. In summer the thermocline separates the warm productive oxygen-rich water of the epilimnion from the cold, unproductive oxygen-deficient hypolimnion. In autumn the epilimnion cools and the aerated water circulates into the hypolimnion as the thermocline disappears.

poorly aerated water below the thermocline is in the *hypolimnion*. The water in the epilimnion and hypolimnion does not mix so that the lake is *stratified*. Most algal productivity occurs in the epilimnion. The algal cells that drop below the thermocline into the hypolimnion are degraded by the heterotrophic microflora. This respiratory activity further depletes the oxygen level in the hypolimnion. In nutrient-rich lakes the hypolimnion is usually totally deficient in oxygen by midsummer.

The temperature of the surface epilimnion water declines in the autumn giving rise to an *overturn*. This is a result of renewed circulation between the epilimnion and hypolimnion in the absence of a temperature difference between them. The overturn reaerates the hypolimnion.

In cold climates winter stratification occurs. In this case cold water is on the surface, trapping a zone of warm water below. The spring overturn occurs when the epilimnion warms to the same temperature as the hypolimnion. This overturn releases nutrients to be used later in the season for algal growth as well as for reaerating the bottom waters.

This picture is of great importance in highly eutrophic lakes. By the middle of August the massive algal biomass begins to decline, depleting the oxygen in the hypolimnion. The bottom waters are used by the fish for spawning. Oxygen deficiency prevents reproduction of many lake fish. This situation has developed in many lakes.

The Sea   The oceans and seas of the world can be divided into two zones:

1. The open sea, which has a very low nutrient concentration. The algae are predominantly phytoplankton. An exception is the Saragasso Sea, which is covered with floating mats of the seaweed *Sargassum* (Fig. 11.3). The diatoms and dino-

FIGURE 11.3   A mat of the seaweed *Sargassum* floating on the Sargasso Sea. (Courtesy Massachusetts Audubon Society.)

flagellates are the most common unicellular algae in the sea. It is often suggested that we can farm the oceans for either fish or plankton to increase the world's food supply. This is a misconception. The productivity of the open ocean is far too low to yield any great increase in fish and it seems unlikely that the sparse phytoplankton population could be easily utilized.

2. Coastal waters provide the most productive areas of the sea. This is particularly true of the *upwelling* zones. These are deep ocean waters close to the shore where nutrients are brought from the ocean depths to the surface by movement of the surface water away from the coast. The upwelling areas off the coast of Peru yield more than 10 million lb of fish per year. Almost the entire catch is anchovies.

In shallow coastal regions large quantities of *benthic* algae accumulate. These attach to sediments and rocks on the bottom and are usually multicellular seaweeds. Green benthic algae include the sea lettuce *Ulva* and *Enteromorpha*. They are found in nutrient-rich harbors and estuaries in cold northern waters. The brown alga *Macrocystis* forms huge fields of seaweed off the coast of southern California, providing a breeding ground for a rich and diverse fish population.

MEASUREMENT OF PRODUCTIVITY

The growth of algae is dependent on the fixation of light energy by the chlorophylls to yield cell mass and oxygen according to the reaction

$$CO_2 + H_2O \xrightarrow[\text{chlorophylls}]{\text{light}} \text{biomass } (CH_2O)_n + O_2$$

Four methods of measuring algal growth are described:

1. biomass determination,
2. oxygen evolution,
3. carbon-14 uptake,
4. DNA and ATP concentration.

We must first differentiate between biomass or standing crop and primary productivity. The *biomass* is the amount of plant material present at any time.

Biomass = amount produced – (amount grazed by higher organisms + amount decomposed by micro-organisms)

The *primary productivity* is the rate at which plant material is synthesized. It is a direct measurement of nutrient and light utilization. Methods 2 to 4 are for measurement of productivity.

Biomass   The simplest method of measuring the output of a natural water is to follow the change in biomass. The dry weight may be determined by sampling throughout the season. Alternatively the maximal possible biomass may be determined by inoculation of the dominant alga to a sample of the water in the laboratory. Cell count or total chlorophyll can be substituted for dry weight measurement. The determination of rate of productivity yields a far more accurate picture of the physical and chemical factors controlling algal growth since we are measuring the rate of photosynthesis.

Dissolved Oxygen   Oxygen evolution is the simplest parameter utilized to measure productivity. One mole of oxygen is released for each mole of CO_2 fixed by photosynthesis. The rate of oxygen evolution is determined by measuring the change in dissolved oxygen concentration in the water. A sample of water is placed in a bottle, inoculated with the algae, and placed back in the water at a specific location or in the laboratory at a specific light and temperature. This procedure is known as a *light bottle test*. Since the algae usually have a microbial population associated with them, there is significant oxygen utilization by microbial respiration. This is measured and corrected for by placing a dark bottle in parallel with the light bottle.

Dissolved oxygen is measured chemically by the Winkler technique (p. 138). The membrane electrode is a more recent innovation for measurement of dissolved oxygen. The technique is based on the detection of diffusion of oxygen across a membrane. The change in oxygen concentration across the membrane is recorded potentiometrically. A dissolved oxygen analyzer is shown on p. 95. Continuous monitoring of dissolved oxygen in a number

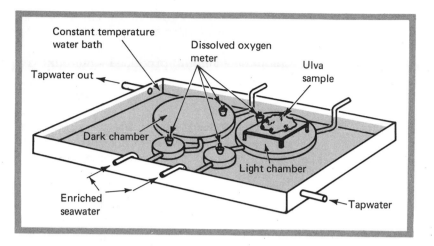

FIGURE 11.4 A schematic drawing of an apparatus used to measure productivity of the benthic alga *Ulva*. One sample is held in the light. A second sample is placed in a dark chamber to correct for dark respiration. Productivity is assayed by measurement of changes in the dissolved oxygen concentration of the water. (Reprinted with permission from T. Waite, L. A. Spielman, and R. Mitchell, *Envir. Sci. Tech.,* 6:1096 (1972). Copyright by the American Chemical Society.)

of different samples can be achieved using this method. Figure 11.4 shows this apparatus being used to determine productivity rates for the benthic alga *Ulva lactuca.*

Carbon-14 Uptake This technique is based on the stoichiometric relationship between CO_2 assimilation and cell synthesis. One mole of reduced organic compound is produced for each mole of CO_2 utilized. The CO_2 uptake is measured by supplementing the CO_2 in the water with radioactive $^{14}CO_2$ supplied as radioactive sodium bicarbonate ($NaH^{14}CO_3$). After a period of hours, or even minutes in highly productive waters, sufficient carbon-14 has been assimilated by the cells to measure emission of the beta radiation in a scintillation counter.

Algae respire and utilize oxygen in the light and in the dark. A bacteria-free algal culture placed in the dark will yield the rate of dark oxygen utilization. Unfortunately, we do not know if the respiration rate is the same in the light. The use of dissolved oxygen for productivity measurement may have an inherent error because of our inability to determine light respiration of algae accurately. Another drawback is the assumption that the algae are only consuming inorganic CO_2. Uptake of organic carbon, if it occurs, is not measured.

DNA and ATP Measurement Only brief mention will be made of these new techniques. They serve to illustrate attempts at more accurate determination of productivity based on essential cell components. The rate of DNA and ATP synthesis directly mirrors the rate of cell productivity. In pure cultures these measurements are extremely sensitive. In natural waters, however, contamination with bacteria and other microorganisms causes serious errors.

It is apparent from this discussion that no single technique yields accurate measurements of algal productivity. Each method has positive and negative attributes. Choice of technique is largely based on arbitrary criteria.

ALGAL NUTRITION

Algae are identical to all other plants in their requirement for inorganic nutrients. The elements required in high concentration are carbon, phosphorus, nitrogen, hydrogen, and oxygen. Carbon must be mainly in the form of CO_2; phosphorus as phosphate; nitrogen as ammonium or nitrate; and hydrogen and oxygen are provided in water. These elements are termed *macronutrients*. When inorganic nitrogen in the water becomes limiting, the nitrogen-fixing blue-green algae often proliferate.

In addition, the algae require low concentrations of many other elements. These include magnesium, sulfur, chlorine, iron, and many metals. These elements are usually present in water in abundant quantities for algal growth. They are called *micronutrients*.

Many algae are incapable of synthesizing some essential metabolites. These materials, called *growth factors*, are added to the medium when algae are grown in the laboratory. In nature, growth factors must be supplied by excretion or degradation products from other organisms. Most marine diatoms require vitamin B_{12} for synthesis. There are abundant quantities of this vitamin in the sea, produced and excreted by bacteria.

ENVIRONMENTAL FACTORS

Light The intensity of light affects both the species of algae in the water and the rate of photosynthesis. Most phytoplankton increase their rate of photosynthesis to an intensity of between 1000 and 1500 lumens (Fig. 11.5). The dinoflagellates are sensitive to intensities above 2000 lumens, and the green algae require almost 800 lumens before growth occurs.

The lower limit of the euphotic zone, or *compensation depth*,

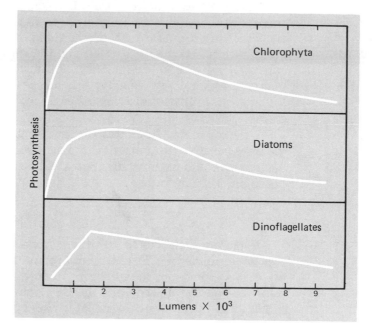

FIGURE 11.5 Relationship between photosynthesis and light intensity for some marine algae. The chlorophyta require 800 lumens for growth and grow well to an intensity of more than 3000 lumens. The diatoms display maximal growth to an intensity of 4000 lumens. Dinoflagellates will grow at low light intensities and are quite sensitive to high intensities. (From J. Ryther, *Limnology and Oceanography*, **1**:72 (1956).)

is between 300 and 500 lumens. The depth of this zone is dependent on the transparency of the water. It may vary from 2 to 3 m in very turbid lakes or harbors to 150 m in the Sargasso Sea.

Temperature   One can find algae growing in the ice water of the Arctic Ocean and in the hot springs of Yellowstone National Park. For each temperature range one can find a group of algae. They are quite specific to that range. As soon as the temperature changes, a new group of algae becomes dominant. In temperate waters there is a clear distinction between a summer and winter population with some overlap. The diatoms that dominate in the spring and early summer increase their productivity rate until the temperature rises above their tolerance level. They decline and the blue-green algae, which have a higher temperature optimum and tolerance, begin to proliferate. The blue-greens tend to be dominant in the autumn.

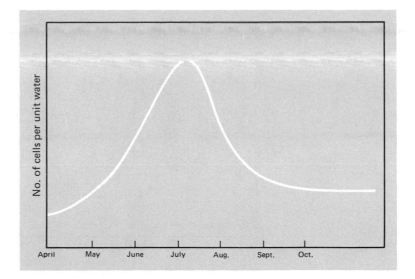

FIGURE 11.6 The increase in population size of the phytoplankton community during the spring and summer.

Factors other than temperature are also involved in species success. It has been suggested that the dominance of blue-green algae is influenced by their toxin production, lowering the numbers of other algae, and by nitrogen deficiency in the water. The decline of diatom blooms is often caused as much by silicon deficiency as by temperature.

The size of the standing crop reflects the water temperature. This is a result of increased productivity rates during warm high light intensity months of the year. The population size of the phytoplankton community changes during the year (Fig. 11.6). Both photosynthesis and grazing rates decline during the cold dark winter months. This picture holds for both marine and freshwater habitats in nontropical climates.

LIMITING FACTORS   Plant growth is controlled by the concentration of nutrients. The yield is directly proportional to the concentration of nutrients which are present in minimal quantities, rather than those nutrients which are abundant. This is stated in Liebig's Law of the Minimum, which was enunciated in the mid-nineteenth century. The law says that the growth of a plant is dependent on the foodstuff presented to it in minimal quantities. Liebig's law is shown schematically in Fig. 11.7. The increase in yield is proportional to the

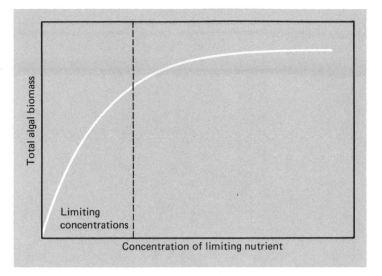

FIGURE 11.7 Schematic drawing showing Liebig's Law of the Minimum. The maximum productivity of an alga is proportional to the concentration of the limiting nutrients. The nutrient is no longer limiting when additional concentrations in the water do not increase productivity.

limiting substrates up to a certain concentration that is dependent on the plant species.

Liebig's law led to the use of fertilizers to increase crop yield and revolutionized agriculture. The law is very pertinent today because of nutrient enrichment of natural waters. We speak of limiting factors in algal productivity. Since nitrogen and phosphorus are most likely to be the least abundant elements in water in relation to algal nutritional requirements, they are usually the elements that determine the size of the standing crop.

The increased concentrations of nitrogen and phosphorus in eutrophic waters are the major sources of excessive algal biomass. If nitrogen or phosphorus are the factors limiting growth of algae, then any increase in their concentration in the water will increase the standing crop of algae. Figure 11.8 depicts the steadily increasing phosphorus concentration in the Zürichsee in Switzerland for the period 1944–1966. Excessive growth of algae began in 1954 and has continued since that time.

This statement describing limiting factors is frequently oversimplified. There is a strong interdependence between nitrogen and phosphorus on the growth of the benthic alga *Ulva* (Fig. 11.9).

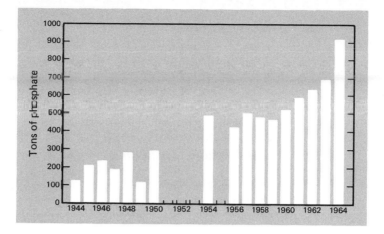

FIGURE 11.8 The phosphorus concentration in a Swiss lake, the Zürichsee. Note the sharp rise in the phosphorous level in the lake after 1954. (From *Eutrophication: Causes, Consequences, Correctives,* Publication ISBN 0-309-01700-9, National Academy of Sciences—National Research Council, Washington, D.C., 1969 (p.38, E.A. Thomas).)

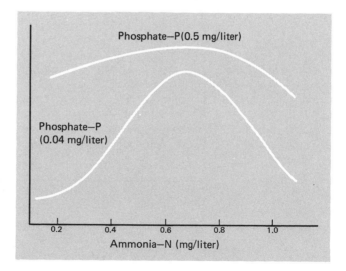

FIGURE 11.9 The synergistic effect of nitrogen and phosphorus on the growth of the benthic alga *Ulva.* When the phosphate concentration is very low (0.04 mg/liter), productivity increases as the ammonia level is increased. At concentrations of NH_3 higher than 0.6 mg/liter, however, productivity is suppressed. When the phosphorus level is increased to 0.5 mg/liter, ammonia is no longer limiting at the concentrations that limited productivity when the phosphorus concentration was low. (From T. Waite, "Ulva Lactuca as a Model for Nutrient Enrichment of Benthic Macrophytes." Ph.D. Thesis, Harvard University, Cambridge, Mass., 1972.)

The response of the alga to nitrogen is much lower at low than at high phosphorus concentrations.

The effect of interactions between environmental factors on species dominance must also be taken into account. The effect of nitrate concentration on growth of four algae is different at two different light intensities (Fig. 11.10). At low light intensity the coccolithophore *Coccolithus* predominates. A totally different re-

FIGURE 11.10 The effect of nitrate concentration on the growth of four algae at two different light levels. Completely different responses to nitrate are observed for each algae at different light intensities. (From R. W. Eppley, et al., *Limnology and Oceanography,* **14:**912 (1969).)

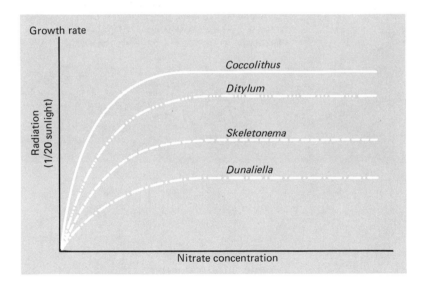

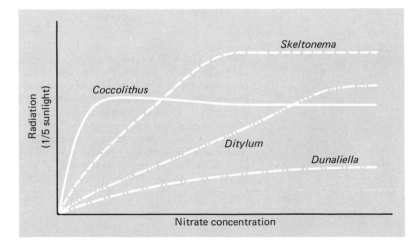

sponse curve occurs at high light intensity. Here the diatom *Skeletonema* predominates, showing a much stronger response to addition of nitrogen.

PHOTOSYNTHESIS AND RESPIRATION

At increasing depths in water the respiration rate is greater than the rate of photosynthesis. The sources of respiration are

1. respiration of the algae,
2. grazing on the algae by zooplankton and fish,
3. biodegradation of algae and organic substrates by micro-organisms.

The photosynthesis:respiration ratio is a very useful tool in the determination of the health of a body of water. In the euphotic zone of oligotrophic waters the photosynthesis:respiration ratio (P:R) is 1. This tells us that respiration by heterotrophic consumers balances the primary productivity.

In nutrient-enriched eutrophic waters, photosynthesis outstrips the ability of consumers to utilize the algae and P:R becomes greater than unity. When a nutrient or light becomes limiting, photosynthesis slows down. There is now an overabundance of substrate for heterotrophic microbial activity. The P:R declines and becomes less than unity. Nutrients are released and a new cycle of intense photosynthesis begins.

These large oscillations between P:R > 1 and P:R < 1 are typical of eutrophic waters. The rich nutrient diet stimulates high algal productivity. This stage is invariably followed by a dramatic decline in productivity. The resultant microbial growth causes serious depletion in the dissolved oxygen in the water.

SOURCES OF NUTRIENTS

Agriculture The farm is the major source of algal nutrients reaching our natural waters. The most important sources are

1. fertilizers and soil organic matter,
2. animal wastes.

Fertilizers often run off agricultural land providing a direct source of nutrients. More than 6 million tons of nitrogen fertilizers and 2 million tons of phosphates are applied to crops in the United States each year. An undetermined quantity is washed by irriga-

FIGURE 11.11 The "Hippo Pool" in Queen Elizabeth Park, Uganda. The lake is heavily eutrophied, probably as a result of excessive use by wildlife. (Courtesy M. V. Williams.)

tion water and rainfall into lakes, streams, and the sea. Fertilizer is inexpensive and overfertilization is common.

The increased crop yields that are a result of modern agricultural practice provide another source of nutrients for natural waters. A good well-managed agricultural soil is rich in organic nitrogen. Microbial degradation of the organic matter releases ammonia and ultimately nitrate. The soluble nitrate salts are leached out and enter either the groundwaters or lakes and streams.

Livestock provide a major source of algal nutrients in the United States. There are more than 600 million livestock in the country at any time. These produce more than 1000 million tons of solid wastes and 500 million tons of liquid wastes. The problem is compounded by the "urbanization" of domestic animals (p. 133).

Eutrophication frequently occurs without human intervention. Figure 11.11 shows a eutrophic lake in Uganda used by animals as a watering hole that is totally covered by water lilies. The nutrient source in this case probably is the excreta of the wildlife using the lake.

Urban Sources The movement of people from rural areas to cities together with the population explosion during the past 50 years have created massive eutrophication problems. Today more than 200 million people live mainly in concentrated metropolitan areas. By 2000 A.D., 77% of the projected 312 million people in

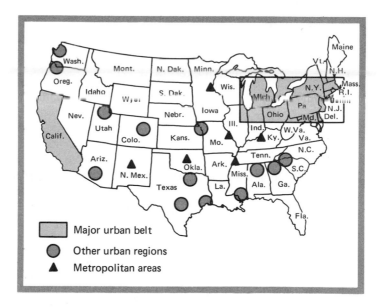

FIGURE 11.12 A projection of the urbanization of the United States for 2000 A. D. By that time there will be 312 million people and 77% will live in the urban and metropolitan areas. (From *Eutrophication: Causes, Consequences, Correctives,* Publication ISBN 0-309-01700-9, National Academy of Sciences—National Research Council, Washington, D.C., 1969.)

the United States are expected to live in the urban areas shown in Fig. 11.12.

Lake Erie provides a model for the contribution of urban wastes to the eutrophication of natural waters. The population along the shores of the lake is now more than 10 million. Four major urban sources of phosphate entering the lake can be recognized (Table 11.1):

TABLE 11.1 Sources of phosphate entering Lake Erie.

Detergents are responsible for more than all other municipal sources combined.

Source	Phosphate (lb/day)
Rural runoff	20,000
Municipal	
Detergents	70,000
Human excreta	30,000
Urban land runoff	6,000
Industrial	6,000
Total	132,000

1. detergents,

2. human excreta,

3. urban land runoff,

4. industrial effluents.

High phosphate detergents provide the major source of phosphates entering natural waters in urban communities today. The recognition that they are major contributors to eutrophication is leading to the development of detergents containing low concentrations of phosphates.

Many cities in the United States have combined sewage and storm drains. During periods of high rainfall or spring thaw in snow zones the untreated sewage is carried with the storm water. Rivers, lakes, and estuaries fed by these combined sewers are usually eutrophied.

Even in cities with separate storm and sewer drains, human excreta are a major source of nutrients. Excreta that are carried to treatment facilities may undergo primary treatment and chlorination. A small percentage of the nutrients are removed in the primary settling process. Secondary treatment removed approximately 30% of the nitrogen and phosphorus. Few communities have the advanced treatment necessary to remove nutrients.

ALGAL TOXINS   Algal blooms constitute an aesthetic and economic problem because of bad odors and fish kills associated with oxygen depletion of the water. In addition, they pose a serious health hazard to man. Many algae produce potent toxins. The dinoflagellate *Ganyaulax catanella* produces *red tides* in the Atlantic and Pacific Oceans (p. 76). The toxin produced by this alga is almost as potent as the extremely poisonous botulism toxin. "Paralytic shellfish poisoning" is the name given to the disease caused by consumption of shellfish contaminated with toxic dinoflagellates. Many deaths have been attributed to consumption of shellfish containing these toxins.

Fish kills have also been associated with *Gonyaulax* and other toxic dinoflagellates. *Prymnesium parvum* is responsible for massive destruction of commercial fish in artificial ponds in many parts of the world. It produces an *endotoxin*, i.e., a poison that is only released following the death of the cell.

Many common blue-green algae also produce endotoxins. The filamentous blue-greens *Anabaena flos-aquae*, *Microcystis aeru-*

ginosa, and *Nostoc* are all toxigenic. These algae are all present and easily isolated from rivers and lakes. Nutrient enrichment of these waters provides the potential for fish kills by the toxins. In addition, there is always a serious danger of contamination of drinking water supplies when excessive quantities of blue-green algae develop. Treatment of natural waters for control of these algae must be carried out early in the season before the biomass becomes too large. Chemical treatment of a large concentration of toxigenic algae would release a massive concentration of endotoxin when the cells lysed.

TASTES AND ODORS   In addition to the production of toxins, blue-green algae have been associated with musty tastes and odors in drinking waters. Water coming from reservoirs containing even small numbers of these algae often has an unpleasant musty smell.

Two blue-green algae, *Symploca muscorum* and *Oscillatoria tenuis*, produce an earthy smelling organic compound that has been given the name "geosmin."

Other blue-green algae that produce tastes and odors include *Anabaena*, *Aphanizomenon*, and *Synura*. The diatom *Asterionella* occasionally is present in drinking waters and produces unpleasant

TABLE 11.2   **Some common chemicals that produce tastes and odors in drinking water.**[a]

Compounds
Mercaptans
Amines
Indoles
Organic acids
Skatoles
Alcohols
Aldehydes
Ketones (e.g., heptanone)
Sulfides
Simple compounds of nitrogens, bromine, chlorine, and sulfur

[a]From F. M. Middleton and A. A. Rosen, *Public Health Reports,* **71**:1125 (1956).

tastes. Actinomycetes growing on organic matter in the water or sediment produce a number of earthy-smelling products that add an unpleasant taste to drinking water at extremely low concentration.

The amount of material necessary to produce a taste or odor in water can be measured in parts per million or even in parts per billion. Some of the more common chemicals producing tastes and odors in drinking water are shown in Table 11.2. The mercaptans are especially odoriferous. A few parts per billion impart a thoroughly unpleasant odor to drinking water. Amines, indoles, skatoles, and organic acids contribute significantly to tastes and odors.

Chemicals producing tastes and odors are eradicated by passing the drinking water through an activated carbon column that adsorbs the organic compounds from the water. The treatment of drinking water is discussed on p. 273.

CONTROL OF EUTROPHICATION

Five methods of controlling eutrophication are available:

1. ecological management,
2. advanced waste treatment,
3. chemical algicides,
4. biological algicides,
5. destratification.

Ecological Management   The most direct means of controlling eutrophication is to limit the flow of nutrients into natural waters. This can be achieved by adequate *ecological management*. The emphasis in this means of treatment should be placed on legislation to control agricultural and urban disposal practices. Excessive fertilization of crops would be forbidden. Feedlots would be required to maintain treatment facilities to remove nutrients. Urban communities would be required to build advanced treatment plants for nutrient removal and industries would be forbidden to dispose of nutrient-rich effluents.

Advanced Treatment   Ecological management is based on the assumption that techniques are available to remove nutrients. Advanced treatment methods are currently in the developmental stage. They are discussed in detail in Chapter 16.

In the absence of current techniques for removing nutrients we must rely on control of excessive algal growth in the presence of the nutrients. These techniques must necessarily be confined to small ponds and reservoirs.

Chemical Algicides Application of copper sulfate is the most common means of preventing excessive algal growth in ponds. It is advisable to add the copper early in the spring before substantial algal growth has occurred. Fish kills frequently occur as a result of the release of intracellular algal toxins or because of microbial

TABLE 11.3 Concentrations of copper sulfate necessary to control algae causing problems in natural waters.[a]

	Organism	Problem	Amount of Copper Sulfate Required (mg/liter)
Algae			
Diatoms	Asterionella, Synedra, Tabellaria	Odor: aromatic to fishy	0.1–0.5
	Fragillaria, Navicula	Turbidity	0.1–0.3
	Melosira	Turbidity	0.2
Greens	Eudorina, Pandorina	Odor: fishy	2–10
	Volvox	Odor: fishy	0.25
	Chara, Cladophora	Turbidity, scum	0.1–0.5
	Coelastrum, Spirogyra	Turbidity, scum	0.1–0.3
Blue-greens	Anabaena, Aphanizomenon	Odor: moldy, grassy, vile	0.1–0.5
	Clathrocystis, Coelosphaerium	Odor: grassy, vile	0.1–0.3
	Oscillatoria	Turbidity	0.2–0.5
Golden or yellow-browns	Cryptomonas	Odor: aromatic	0.2–0.5
	Dinobryon	Odor: aromatic to fishy	0.2
	Mallomonas	Odor: aromatic	0.2–0.5
	Synura	Taste: cucumber	0.1–0.3
	Uroglenopsis	Odor: fishy; taste: oily	0.1–0.2
Dinoflagellates	Ceratium	Odor: fishy, vile	0.2–0.3
	Glenodinium	Odor: fishy	0.2–0.5
	Peridinium	Odor: fishy	0.5–2.0

[a]From G. M. Fair, et al., *Water and Wastewater Engineering*, Vol. 2. John Wiley & Sons, Inc., N. Y., 1968.

growth and consequent oxygen depletion following the release of large quantities of algal organic matter.

Copper sulfate is relatively nontoxic to humans or fish at the concentrations required to kill algae. Concentrations of 0.1–0.5 mg/liter of the water in the photic zone are usually applied. The concentration is much lower when based on the total body of water. Trout are susceptible to 0.14 mg/liter of copper sulfate but are rarely killed by treatment of lakes. The U.S. Public Health Service permits 1 mg/liter in drinking water.

Table 11.3 illustrates the fact that no single group of algae is responsible for tastes and odors and each group required a different concentration of copper sulfate for control. Diatoms, greens, blue-greens, and dinoflagellates all are capable of producing products that taste or smell unpleasant in drinking water. Usually chlorination will prevent growth of these algae; however, the products are usually formed in the reservoir prior to treatment.

Biological Algicides This method is applicable only to lakes and ponds. Algae are highly susceptible to predation by bacteria when they are short of food or are limited by low light intensity or cold temperatures. However, inoculation of bacterial predators to actively growing algae has not been successful in lowering the standing crop.

TABLE 11.4   Biological control of excessive growth of blue-green algae.

Viruses active against blue-green algae were inoculated and drastically reduced the algal population. This method of control has not been tested on a field scale.[a]

Days	Algal Filaments (length/ml, μ)		Virus Count (Plaque Forming) (units/ml)
	0-100	100-300	
0	115,000	20,000	6×10^5
3	165,000	55,000	4×10^4
6	—	—	1×10^7
7	5,000	0	2×10^9

[a]Reprinted from *Journal American Water Works Association*, **56**:1217 by permission of the Association. Copyright 1964 by the American Water Works Association, Inc., 2 Park Avenue, N. Y., 10016

Viruses have been isolated that are active against blue-green algae. These can be used to control algal growth (Table 11.4). The blue-green algal viruses are genus specific. They have not been tested in natural conditions. Use of these viruses for control of blue-green algae would require (1) inoculation into the water of a broad spectrum of viruses active against a number of different algae and (2) inoculation very early in the season to prevent development of excessive algal growth. No viruses are available to control green algae.

Destratification   The stratification of lakes and reservoirs is described on p. 191. Artificial destratification can be used to control eutrophication. Destratification can be achieved by either physical mixing or forced aeration of the lake. Figure 11.13 shows that artificial destratification drastically lowers the standing crop.

Artificial destratification changes the algal dominance in the lake (Fig. 11.14). It causes a decline in the population of blue-green algae and a relative increase in the numbers of green algae.

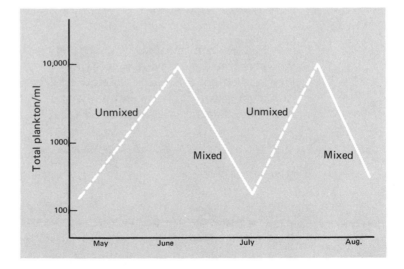

FIGURE 11.13 The effect of artificial destratification on the algal population of a lake. In an undisturbed lake the biomass increases during the summer. Destratification immediately results in a decline in the phytoplankton count. It must be emphasized that mixing is not effective in shallow ponds. (From J. Ridley and J. M. Symons, *Water Pollution Microbiology*, ed. R. Mitchell. John Wiley & Sons, Inc., N. Y., 1972.)

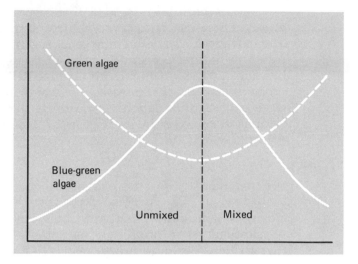

FIGURE 11.14 In an undisturbed lake early in midsummer the blue-green algae are dominant. Artificial destratification causes a decline in the blue-green algal population and a relative increase in the population of green algae. This is probably a result of insufficient light in the mixed lake for blue-greens. (From J. Ridley and J. M. Symons, *Water Pollution Microbiology,* ed. R. Mitchell. John Wiley & Sons, Inc., N. Y., 1972.)

The mechanism of algal control by destratification is unknown; however, evidence points to control by light limitation under destratified conditions. This method is becoming common to control algal growth in lakes and reservoirs. It is usually only effective when the depth is greater than 40 feet.

SUMMARY

1. The predominant primary producers in rivers are algae. In slow streams, a sessile plant population develops.

2. Deep lakes are thermally stratified. Algal growth occurs in the well-aerated epilimnion.

3. The open ocean is a biological desert. Most productivity occurs in the nutrient-rich coastal waters, particularly in upwelling zones.

4. Algal productivity is usually measured by determination of biomass, oxygen evolution, carbon-14 uptake, DNA or ATP determination.

5. Growth of algae is controlled by macronutrients and micronutrients, light, and temperature of the habitat.

6. Excessive concentrations of phosphorus and nitrogen are responsible for eutrophication. Sources are both agricultural and urban.

7. Excessive algal growth causes fish kills as well as tastes and odors in drinking water.

8. Eutrophication can be controlled by ecological management, limiting nutrient inputs, chemical or biological algicides and by destratification in lakes.

FURTHER READING

J. J. Goering, "The Role of Nitrogen in Eutrophic Processes" in *Water Pollution Microbiology*, ed. R. Mitchell. John Wiley & Sons, Inc., New York, N. Y., 1972.

National Academy of Sciences, *Eutrophication: Causes, Consequences, Correctives*, Washington, D. C., 1969.

J. E. Ridley and J. M. Symons, "New Approaches to Water Quality Control in Impoundments" in *Water Pollution Microbiology*, ed. R. Mitchell. John Wiley & Sons, Inc., New York, N. Y., 1972.

W. Stumm and E. Stumm-Zollinger, "The Role of Phosphorus in Eutrophication" in *Water Pollution Microbiology*, ed. R. Mitchell. John Wiley & Sons, Inc., New York, N.Y., 1972.

METALS

AS

POLLUTANTS

12

A vast store of metals is locked in the world's mineral deposits. These metals are released and concentrated in mining operations. The low-grade ores, or *tailings*, are accumulated beside the mine. Tailings are exposed to rainfall and act as a reservoir for metals that are washed into streams. It has been estimated that by 1980 the combined *solid* waste output from these industries will reach 2 billion tons annually. Often the concentration of metals dissolved in a river is sufficiently high to eradicate the fish population or to render it inedible.

Abandoned coal mines are another major source of metal pollution. Water flowing through the mines reacts with iron-bearing pyritic minerals to produce acidity. More than 10,000 miles of waterways in the United States are contaminated by *acid mine drainage*.

Industries release many other metals into the environment. Our natural waters receive large quantities of mercury, lead, nickel, copper, zinc, chromium, and cadmium. It has been assumed that natural waters provide infinite dilution for heavy metals. However, even extremely low concentrations of metals are biomagnified.

CORROSION OF IRON

Iron in pipes buried underground frequently corrodes. When conditions become severely oxygen deficient, bacterial *corrosion cells* form on the pipe. Soils that are poorly drained are particularly susceptible to corrosion. A badly corroded water pipe is shown in Fig. 12.1.

The corrosion process requires the presence of sulfate. The

FIGURE 12.1   A corroded iron pipe. The corrosion cells form pits that ultimately penetrate the full thickness of the pipe.

first step is reduction of sulfate to H_2S by the obligately anaerobic bacterium *Desulfovibrio*. This organism is a short motile rod and it is heterotrophic. Bacterial corrosion does not occur in the absence of available organic matter.

The second step in the reaction is the reduction of elemental iron to Fe^{2+}. This reaction is autocatalytic under reducing conditions.

Finally, the reduced iron reacts with hydrogen sulfide and water to yield ferrous sulfide and ferrous hydrate:

$$4\,Fe^{2+} + SO_4^{2-} + 4\,H_2O \rightarrow FeS + 3\,Fe(OH)_2 + 2\,OH^-$$

These products are washed away, leaving a pit in the pipe. Bacterial corrosion only occurs when the temperature is between 10 and 30°C and at pH values above 5.5. These conditions are delineated by the growth characteristics of *Desulfovibrio*.

ACID MINE DRAINAGE

Drainage water from abandoned mines is acid and rich in iron oxides. The water contaminates streams, killing fish and destroying drinking water supplies. Iron oxide precipitates discolor the water and destroy recreational areas.

The rate-limiting step in the production of acidity is the oxidation of ferrous iron from pyrites in the mine to ferric iron. It is initiated by the decomposition of solid iron pyrites (ferrous sulfide) releasing soluble ferrous iron:

$$(1) \quad FeS_2 + 7/2\, O_2 + H_2O \rightarrow Fe^{2+} + 2\, SO_4^{2-} + 2\, H^+$$

The soluble ferrous iron is oxidized in the presence of oxygen to ferric:

$$(2) \quad Fe^{2+} + 1/4\, O_2 + H^+ \rightarrow Fe^{3+} + 1/2\, H_2O$$

The Fe^{3+} may be precipitated:

$$Fe^{3+} + 3\, H_2O = Fe(OH)_3 + 3\, H^+$$

or it may react with pyrite to yield more Fe^{3+}:

$$(3) \quad FeS_2 + 14\, Fe^{3+} + 8\, H_2O \rightarrow 15\, Fe^{3+} + 2\, SO_4^{2-} + 16\, H^+$$

A cycle of release of ferrous iron from pyrites and oxidation to ferric iron is developed with ferric iron acting as the catalyst to drive reaction 3 and close the cycle.

The sulfur portion of the pyrites reacts chemically with Fe^{3+} and with O_2 to produce more acidity:

$$(4) \quad S + 6\, Fe^{3+} + 4\, H_2O \rightarrow 6\, Fe^{2+} + SO_4^{2-} + 8\, H^+$$

$$(5) \quad S + 3/2\, O_2 + H_2O \rightarrow SO_4^{2-} + 2\, H^+$$

The Iron-Oxidizing Bacteria   The oxidation of ferrous to ferric iron controls the production of acidity in mine drainage waters. When the water pH is above 4.5, oxidation occurs chemically without any microbal mediation. In water with a pH below 4.5, chemical oxidation of ferrous iron is infinitely slow. Under these conditions the iron-oxidizing bacteria are responsible for the reaction oxidizing ferrous to ferric iron. The iron bacteria can be divided morphologically into three groups:

1. *The Haplobacteria*. These are Gram-negative nonspore-forming rods. The most important iron-oxidizing haplo-bacterium is *Thiobacillus ferrooxidans* (Fig. 12.2). *T. ferrooxidans* is an autotroph requiring ferrous iron for growth. It is viable in the pH range from 2.2 to 4.6. Iron oxidation occurs only between pH 2.4 and 3.5. The organism does not become encrusted with iron. Ferric iron precipitates deposit in the water.

   Other haplobacteria occur in pyrite ores. They do not utilize iron as an energy source, however, and are probably

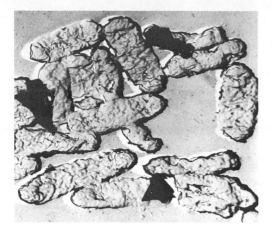

FIGURE 12.2 *Thiobacillus ferrooxidans,* a chemoautotrophic bacterium that requires ferrous iron for growth. (Courtesy D. Lundgren.)

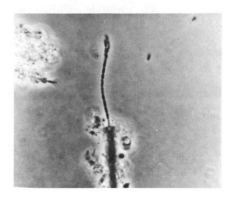

FIGURE 12.3 Iron encrusted sheath of the stalked bacterium *Leptothrix.* A chain of cells is growing from the sheath. Magnification 1200X. (Courtesy P. Hirsch.)

FIGURE 12.4 An electron micrograph of the iron-oxidizing stalked bacterium *Gallionella* showing the cell and secreted stalk of ferric hydroxide. (Courtesy R. Wolfe.)

of minor importance in iron oxidation. They include *Sidercapsa*, a bacterium that forms chains of cocci embedded in a capsule, and *Naumanniella*, a unicellular rod in which iron is deposited in the form of a solid ring.

2. *The Sheathed Bacteria. Sphaerotilus,* (p. 142), is the most common bacterium of this group. The organism forms chains of rods held in a sheath. The pH range is 5.8 to 8.5. At these pH values, rapid chemical oxidation of iron occurs and biological oxidation is not ecologically significant. Growth is heterotrophic. *Leptothrix* (Fig. 12.3) and *Crenothrix* are morphological variants of *Sphaerotilus*.

3. *Stalked Bacteria. Gallionella ferruginea* is the most common of the stalked iron bacteria. It has a bean-shaped cell with a nonliving stalk encrusted with iron hanging from it (Fig. 12.4). The organism grows well heterotrophically and

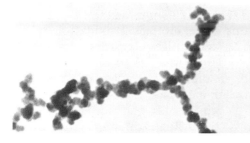

FIGURE 12.5 An electron micrograph of an iron-oxidizing strain of *Metallogenium*. The stalks are encrusted with ferric hydroxide. (Reprinted with permission from F. Walsh and R. Mitchell, *Envir. Sci. Tech.,* 6:809 (1972). Copyright by the American Chemical Society.)

has a pH optimum of 6.0 so probably it is not important in iron oxidation processes in nature.

Metallogenium is an iron-oxidizing stalked bacterium without a distinct cell body. The organism forms intertwining masses of filaments heavily encrusted with ferric iron. A typical growth of *Metallogenium* is shown in Fig. 12.5. The bacterium has a pH range of 4.0 to 6.8.

Ecological Succession   The production of acidity in mine drainage waters is dependent on the chemical and biological oxidation of iron, which in turn depends on an ecological succession shown in Fig. 12.6. When neutral pH water passes through pyritic minerals, chemical oxidation accounts for the formation of ferric ions and

FIGURE 12.6   The ecological succession leading to the development of acid mine waters.

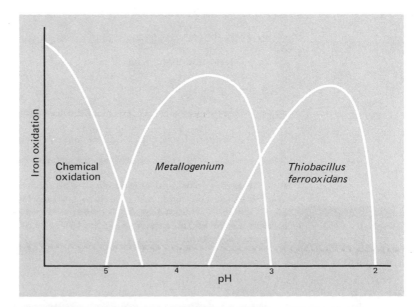

the consequent development of acidity. Chemical catalysis becomes insignificant at pH 4.5. At this stage biological oxidation takes over. The stalked bacterium *Metallogenium* becomes dominant and catalyzes iron oxidation, reducing the pH to 3.5. This pH is toxic to *Metallogenium* and it dies out to be replaced by *Thiobacillus ferrooxidans*. Further catalysis of iron oxidation by this bacterium brings the pH to 2.5. No further biological oxidation occurs and this is the pH of the effluent drainage water from the mine. The succession of chemical and biological processes reduces the pH of the mine water from 7.0 in influent water to 2.5 in the drainage water.

Control   Ecological management of mines may provide the solution to the acidity problem. Some preventive measures are summarized in Fig. 12.7. Abandoned mines have been sealed to prevent access of oxygen which is essential for iron oxidation. This is technically difficult and has met with little success. Prevention of water flow through the mine is usually not feasible. The use of chemicals to kill iron bacteria is unacceptable because of contamination of the water with toxic chemicals. It may be possible to achieve biological control by developing a microflora antagonistic to *Metallogenium*. Alternatively, the ecological succession could be broken by inhibiting *Metallogenium* with high concentrations of ferric iron. This would prevent the pH from dropping below the 4.5 produced by chemical oxidation of ferrous

---

Possible Prevention of Acid Mine Drainage:

1.   Seal abandoned mines.

2.   Prevent flow through mines.

3.   Chemical kill of iron bacteria.

4.   Development of an antagonistic microflora against *Metallogenium*.

5.   Inhibition of growth of *Metallogenium* by excess ferric iron.

FIGURE 12.7   Some possible solutions to the problem of acid mine drainage. The most promising solution is offered by breaking the ecological succession. *Metallogenium* is inhibited by high concentrations of ferric iron. Recycling oxidized wastes can prevent the pH from falling below pH 4.5 by inhibiting *Metallogenium* growth.

iron. Recycling of drainage water containing ferric iron through the mine could be used to break the succession.

MANGANESE

Manganese, like iron, exists in the soluble reduced form in oxygen-deficient waters. Drinking water containing iron and manganese often has a brown coloration. Unlike iron, reduced manganous ion (Mn^{2+}) is chemically stable at neutral pH. It is precipitated by lime in water treatment plants as manganous hydroxide [$Mn(OH)_2$].

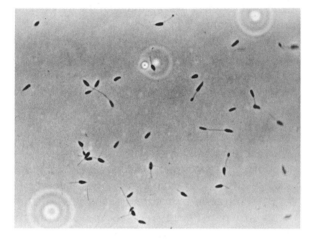

FIGURE 12.8 A photomicrograph of a mat of *Hypho-microbium.* (Courtesy P. Hirsch.)

Manganese-oxidizing bacteria are important in hydroelectric pipes and possibly in domestic water pipes where manganese encrustations occur. Deposits of manganic oxides on hydroelectric pipes cause severe loss in water pressure. The oxidation of soluble Mn^{2+} to insoluble MnO_2 on surfaces in neutral pH waters is catalyzed by a stalked bacterium that attaches to the surface. The stalks trail in the water acting as a form of sponge for the Mn^{2+}.

The reaction $Mn^{2+} \rightarrow Mn^{4+}$ occurs on the stalk surface. Layers of MnO_4 form on the pipe surfaces and are usually enmeshed in a mat of stalks. Figure 12.8 shows a mat of the stalked manganese oxidizing bacterium *Hyphomicrobium.* An electron micrograph of a pure culture of *Hyphomicrobium* is seen in

FIGURE 12.9 An electron micrograph of *Hyphomicrob-ium.* (Courtesy K. C. Marshall.)

Fig. 12.9. Swarm cells swim freely in the water. These attach to the surface and develop stalks. The stalks branch and yield budding cells at irregular intervals.

The stalked bacterium *Metallogenium* that is involved in iron oxidation is also found in manganese oxide deposits. Pure cultures of *Metallogenium* capable of oxidizing manganese have been isolated.

The mechanism of manganese oxidation by bacteria has not been elucidated. It is possible that the organisms are chemoauto-trophic and utilize the manganese as an energy source. Alternatively the bacteria may utilize an organic energy source. An increase in the pH at the surface of the stalk may cause precipitation of manganese oxides.

No control measures have been devised. Hydroelectric systems are periodically closed for cleaning of manganese encrustations. Water mains are flushed at high pressure to dislodge iron and manganese precipitates on their surface.

MERCURY   Mercury contamination of the environment is hazardous to both humans and wildlife. The first well-documented outbreak of waterborne mercury poisoning occurred in Minamata in Japan in 1953. People eating fish or shellfish from Minamata Bay became ill with "Minamata disease," which attacked the central nervous system. The disease was traced to the presence of methyl mercury in the fish. Inorganic mercury was being discharged into the bay by the chemical industry. Since that time accumulations of methyl mercury in fish have been documented from different parts of the world.

Sources   Mercury is widely used in agriculture and in industry. The major sources are:

1. organomercury pesticides used in agriculture,

2. mercury slimicides used in the paper industry,

3. mercury paints,

4. chlorine manufacture.

Plant seed treatment with mercury compounds is very common as a means of protecting the seed against fungal deterioration. Mercury paints provide strong protection against fouling of boats, although their use is being discontinued. Phenylmercuric acetate has been used for many years to prevent bacterial slime formation in the production of paper. Sediments close to paper plants still contain large quantities of mercury despite use of alternative non-mercury bactericides.

The United States manufactures thousands of tons of chlorine each year. At least 25% is produced by electrolysis of sodium chloride using mercury electrodes. The waste mercury from these electrodes is disposed into natural waters. The St. Clair River that runs between Michigan and Ontario has absorbed 20,000 tons of mercury during the past 20 years.

In addition to these sources, broken thermometers, electrical equipment, and other consumer products contribute a significant but unknown quantity of mercury to the environment.

Metabolism   Both metallic mercury and many organic mercury compounds are metabolized by microorganisms. Aerobic bacteria convert metallic mercury through mercurous ion to either methyl mercury

$$Hg^\circ \rightarrow Hg^{2+} \rightarrow CH_3Hg$$

or to dimethyl mercury

$$Hg^\circ \rightarrow Hg^{2+} \rightarrow CH_3HgCH_3$$

Dimethyl mercury is converted to monomethyl mercury under acidic conditions. The reactions can be seen in Fig. 12.10.

Conversion of other organic mercury compounds to methyl and dimethyl mercury depends on their bacterial conversion to mercurous or mercuric ion. Demethylation by anaerobic bacteria occurs in anoxic sediments.

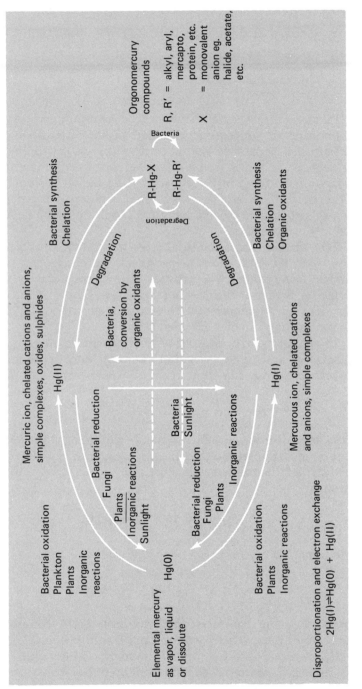

FIGURE 12.10 The reactions controlling the formation of methyl mercury compounds in water. Methyl mercury and dimethyl mercury are formed by bacterial synthesis either from mercuric or mercurous ions. These ions originate as elemental mercury of organomercury compounds. (From I. R. Jonasson and R. W. Boyle, *Mercury in Man's Environment*, Proc. Symp. Royal Society of Canada, Ottawa, Ontario, 1971.)

Biomagnification We are concerned primarily with the *alkyl-mercury* compounds methyl mercury and dimethyl mercury for the following reasons:

1. The compounds are highly soluble in water.
2. They are rapidly and easily absorbed through biological membranes.
3. They are not easily degraded in or released from the animal body.

Alkylmercury compounds are absorbed at the bottom of the food chain by phytoplankton and zooplankton and biomagnified (Fig. 12.11). Swordfish and tuna may contain as much as 0.1 ppm alkylmercury if they are caught near contaminated waters. Mercury in the soil and in the atmosphere is absorbed by plants. The mercury concentration in the marine biota is illustrated in Table 12.1. The littoral flora and fauna in the Tay River estuary

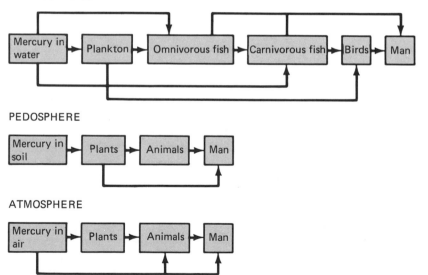

FIGURE 12.11 The biomagnification of methyl mercury compounds. Mercury is biomagnified in water through the primary producers to the fish and birds. The biomagnification process is shortened in soil and in the air where the mercury is transferred directly from plants to animals and ultimately to humans. (From I. R. Jonasson and R. W. Boyle, *Mercury in Man's Environment,* Proc. Symp. Royal Society of Canada, Ottawa, Ontario, 1971.)

TABLE 12.1 The concentration of mercury in animals and plants in the estuary of the River Tay, England.[a]

	µg Hg per Gram Tissue
Seaweeds	
Ulva	25
Fucus	1.1
Laminaria	0.7
Molluscs	
Mytilus	2.1
Littorina	0.1–1.8
Vertebrates	
Grey seal liver	224.8
Eider duck liver	2.1

[a]From A. M. Jones, Y. Jones, and W. D. P. Stewart, *Nature*, **238**:164 (1972).

in England contain significant concentrations of mercury. *Ulva* contained 25µg/g, presumably concentrated by direct absorption from the water. The grey seal in the study contained 224 µg Hg per gram in its liver. It is impossible to tell if these extraordinary high concentrations are obtained by direct uptake or by passage up the food chain.

Mercury toxicity in man initially causes tremors and fatigue. Severe poisoning destroys the brain cells. The accumulation of mercury in the human body depends on (1) rate of consumption, and (2) rate of excretion.

Methyl mercury has a half-life in humans of 70 days. The United States has set a permissible level of methylmercury intake of 0.075 mg/day.

Control   Measures to control poisoning of wildlife and humans by alkylmercury compounds center around the prevention of disposal of both inorganic and organic mercury into natural waters and soil. Use of mercury paints and fungicides is being discontinued in many countries. Laws are being passed requiring recovery of mercury wastes by industry. Concentrated sources of methyl mercury in sediments, such as those that caused Minamata disease, will ultimately disappear. Diffuse sources of mercury, however, including fossil fuels, continue to provide low levels of mercury pollution.

TRACE METALS   It is becoming apparent that atmospheric emissions of trace elements from industries and automobiles are concentrating in the world's oceans. These elements are enriched in marine food chains. The enrichment factor in the biota may be defined as follows:

$$\text{Enrichment factor} = \frac{\mu g \text{ element per g organism}}{\mu g \text{ element per g seawater}}$$

The enrichment of some of the common trace element pollutants in shellfish is shown in Table 12.2. The enrichment of cadmium, copper, nickel, and zinc is striking because these elements either are required in infinitesimally low concentrations by the marine biota or are not required for biological metabolism.

TABLE 12.2   Concentration of trace elements from the ocean into scallops. Similar concentrations occur in oysters and mussels.[a]

Trace Element	Concentration Factor
Cadmium	10^6
Chromium	10^5
Zinc	10^4
Nickel	10^4
Lead	10^3
Copper	10^3
Silver	10^3

[a]From R. R. Brooks and M. G. Runsky, *Limnology and Oceanography,* **10**:521 (1965).

Cadmium   Large quantities of cadmium are used in modern technology. The element is concentrated and magnified in the food chain and is quite toxic. It must be considered as a potentially hazardous metal.

Most of the cadmium polluting the environment originates in atmospheric emissions. Smelting operations are responsible for yielding significant quantities. Cadmium is frequently used as a gasoline additive and is emitted in automobile exhausts. Ambient air concentrations approximate 0.001 μg/cu m although they may be as high as 0.01 over industrial areas. Atmospheric cadmium ultimately accumulates on soil and water when it is washed out in rain. Liquid discharges originate primarily in the mining and electroplating industry.

Cadmium is absorbed directly in the elemental form by plankton and higher organisms. No bacterial metabolism in the natural habitat has been detected. The ambient concentration in the ocean is normally 0.2 ppb. The digestive glands of mollusks may contain as much as 10 ppm of cadmium.

Human toxicity of cadmium is associated with kidney, cardiovascular, and respiratory disease. Normal humans accumulate about 50 mg of cadmium in a lifetime, mainly in the kidneys. The half-life is approximately 10 years. Tolerance levels are only available for drinking water (10 ppb) and air (100 μg/cu m). We know very little about the accumulation of cadmium in our food chains and the long-range toxic effects. In mining areas both agricultural crops and fish have high concentrations of cadmium and increase the daily intake. The extreme stability of this metal in the animal and human body and its concentration in the food chain suggest that emission to air and water should be controlled.

Other Trace Metals   Cadmium provides us with a model for metals that are used extensively in our society and are not metabolized in large quantities by microorganisms. Many of these metals are readily absorbed by plankton, fish, and agricultural crops. High concentrations are absorbed by animals at the top of the food chain and particularly by humans. Lead, zinc, and copper are found in high concentrations in natural waters, particularly close to mining or electroplating operations.

Lead accumulates in bacterial and algal cells without killing the cell. As much as 400 mg of lead per gram of bacterial mass can be immobilized from water rich in soluble lead salts, although such high concentrations are rarely found.

Lead is known to accumulate in submerged plants in the heavily polluted River Ruhr. Both spermatophytes and bryophytes accumulate the metals in concentrations as high as 2000 ppm. Similar accumulations of copper, zinc, and nickel are found in the Ruhr. Any accumulation of lead or other heavy metals in the food chain above those normally found in unpolluted waters must be considered to be hazardous.

Standards   The United States Public Health Service provides standards for trace metals in drinking waters. These are sum-

TABLE 12.3 United States Public Health Service standards for chemical contamination of drinking water.[a]

Metal	Suggested Limit (mg/liter)	Unacceptable (mg/liter)
Arsenic	0.01	0.05
Barium	—	1.0
Cadmium	—	0.01
Chromium (hexavalent)	—	0.05
Copper	1.0	—
Iron	0.3	—
Lead	—	0.05
Manganese	0.05	—
Selenium	—	0.01
Silver	—	0.05
Zinc	5.0	—

[a]From *Standard Methods for the Examination of Water and Wastewater*, 13th ed., American Public Health Association, Washington, D.C., 1971.

marized in Table 12.3. Suggested maximum concentrations are given for the nontoxic metals iron and manganese: 0.3 mg/liter for iron and 0.05 for manganese. The suggested limit for copper is 1.0 mg/liter and for zinc, 5.0 mg/liter.

Drinking waters are rejected as hazardous to health when the concentration of arsenic, lead, chromium, or silver exceeds 0.05 mg/liter or when the selenium or cadmium concentration exceeds 0.01 mg/liter. Barium is permitted to a concentration of 1.0 mg/liter.

RADIONUCLIDES

Sources The development of nuclear power brought with it the dissemination of artificially produced radioactive elements or *radionuclides*. For the past 20 years the major source of radionuclides in the biosphere has been atmospheric testing of nuclear weapons. This source had declined in recent years. It is being replaced at an accelerating rate, however, by the production of nuclear power plants.

These power plants have the potential to produce both long- and short-lived radionuclides in the air or cooling water. Irradiation of elements present in the atmosphere or water as impurities produces the radionuclides. Most of these elements are removed by ion exchange before the air or cooling water is released. Small

TABLE 12.4 Some long-lived radionuclides released by nuclear reactors.

Krypton-85 is released in the atmospheric emissions. All the other elements are released into the cooling water.[a]

Radionuclide	Half-Life
Manganese-54	314 days
Cobalt-58	71 days
Cesium-134	2 years
Cesium-137	30 years
Krypton-85	10 years

[a]From T. R. Rice and D. A. Wolfe, *Impingement of Man on the Oceans*, ed. D. Hood. Wiley-Interscience, N. Y., 1971.

quantities are not removed and are disposed into the environment. Table 12.4 illustrates the major long-lived elements released from a 10,000-megawatt nuclear power plant. The short-lived radionuclides pose a minimum threat. However, long-lived radionuclides such as manganese-56 with a half-life of almost a year, krypton-85 with a half-life of more than 10 years, and cesium-37 with a half-life of 30 years accumulate to high concentrations in the environment.

In addition to nuclear reactors, artificial radionuclides are deliberately disposed into natural waters by hospitals and research laboratories. More than 3 million curies of radioactive chemicals were used for medical and other scientific research in the United States in 1967. A significant portion of this material was disposed in very dilute solutions to wastewaters.

Accumulation by the Biota   Long-lived radionuclides ultimately find their way to the oceans from freshwater runoff and from the atmosphere. There they may remain in solution or are precipitated chemically and fixed in the sediments. Those radioactive elements that remain in solution have the potential to be accumulated in the food chain. Manganese-54, cobalt-58, and cesium-137 are significantly accumulated by many marine organisms (Table 12.5). Most edible fish are caught close to estuaries where there is potentially a continuous source of radionuclides. We know very little about the accumulation by the biota of other radionuclides released from nuclear power plants. No significant uptake of krypton 85 has been observed.

TABLE 12.5   The accumulation of long-life radioactive elements in the food chain.[a]

Radionuclide	Concentration Factor			
	Algae	Crustacea	Molluses	Fish
Manganese-54	3000	2000	10^4	200
Cobalt-58	500	500	500	80
Cesium-137	15	20	10	10
Krypton-85	~1	~1	~1	~1

[a]From T. R. Rice and D. A. Wolfe, *Impingement of Man on the Oceans,* ed. D. Hood. Wiley-Interscience, N. Y., 1971.

Effects of Irradiation   It seems unlikely that sufficient concentrations of radionuclides would accumulate to cause direct toxicity either to marine organisms or to humans. However, continuous exposure of eggs and larvae to significant levels of radioactivity resulting from biomagnification of radionuclides may ultimately cause genetic aberrations. In addition, fertility is affected adversely by radioactivity and the population of organisms concentrating radionuclides may begin to decline.

The threat to humans from radionuclides in power plant cooling water would come from the consumption of seafood. The concentrations could be sufficiently high to cause adverse effects. We know that radionuclides are carcinogenic and may cause deleterious genetic changes. The permissible accumulation in the body should not exceed the natural background level until we know more about hazardous concentrations of radionuclides.

Control   The most effective means of controlling the accumulation of radionuclides in the food chain is to prevent any release into either receiving waters or the atmosphere. This can be achieved by the imposition of more rigorous standards for nuclear power plant effluents and emissions. The use of more powerful ion-exchange systems can yield cooling water and atmospheric emissions free of radionuclides.

SUMMARY

1. Iron pipes corrode under reducing conditions. Pyritic minerals oxidize in mines and form acid mine drainage. Control of mine water acidity may be achieved by breaking the ecological succession of microorganisms.

2. Manganese precipitation occurs in hydroelectric pipes through bacterial manganese oxidation. The predominant bacterium is *Hyphomicrobium.*

3. Mercury presents a hazard in natural waters because of the microbial formation of methyl and dimethyl mercury. These alkyl mercury compounds are biomagnified in the food chain. They are highly toxic to humans and are not rapidly excreted.

4. Cadmium and other trace metals are biomagnified in a similar manner to mercury. They are absorbed directly in the elemental form.

5. Radionuclides are disposed into water from nuclear power plants, hospitals, and research laboratories. They are concentrated in the biota, providing a hazard both to humans and to the aquatic fauna.

FURTHER READING

T. R. Rice and D. Wolfe, "Radioactivity—Chemical and Biological Aspects," in *Impingement of Man on the Oceans* by D. Hood. Wiley-Interscience, New York, N. Y., 1971.

Royal Society of Canada, "Mercury in Man's Environment," Proceedings of a Special Symposium, Ottawa, Ontario, 1971.

J. M. Wood, "Environmental Pollution by Mercury," in *Adv. in Envir. Sci. Tech.*, II (1971).

COMMUNITY
ECOLOGY

13

The myriad of species of microorganisms living together in an ecosystem constitutes a *community*. The microbial community is broad and diverse. A typical sample of soil or water may have as many as 200 different species and many more biochemical mutants. This diversity allows the community to adapt rapidly to changing environmental conditions. While some microorganisms may be sensitive to a specific form of stress, e.g., increased temperature, other microorganisms will thrive under these conditions and fill the niche.

The stability of microbial communities is dependent on this broad biochemical diversity, which is maintained by a complex series of interactions between microorganisms within the ecosystem. No single organism or even group of organisms is allowed to dominate. Chemicals produced by one group of organisms may either nourish or kill another group. These interactions can be categorized as *benevolent* or *antagonistic*.

BENEVOLENT INTERACTIONS

Commensalism A common interaction between two microorganisms involves the production of a substance by one organism that benefits the second. The relationship is called *commensalism*. Two types of commensal interactions can be seen in natural habitats. In one relationship a material is predigested by one organism for use by another. Cellulose decomposition provides a good example. Cellulolytic microorganisms, such as *Cellulomonas*, degrade the cellulose to yield glucose, which is utilized as a substrate by saccharolytic microorganisms of which yeasts are typical. Many microorganisms cannot synthesize one or more metabolites essential for growth. These metabolites are then termed *vitamins* or *growth* *factors.* Yeasts produce a number of different growth factors including B vitamins, which many bacteria are incapable of synthesizing. Aerobic and anaerobic microorganisms can grow in the same niche because the aerobes deplete the oxygen and allow the anaerobes to develop.

Protocooperation Frequently growth factors are exchanged between microorganisms. This interaction is called *protocooperation.* An example of this form of mutual benevolence is the interaction between mutants of *Escherichia coli*, each of which has lost the ability to synthesize an essential metabolite. When we place two mutants together, one of which has lost the ability to synthesize ornithine and another which has lost the ability to synthesize arginine, both mutants can grow using the excretion of ornithine or arginine by the other.

Symbiosis, the most extreme form of mutually benevolent interaction, involves an extremely close physical association and interchange of physiological functions. The *lichen* is a prime example. Lichens are associations between an alga and a fungus. The alga may belong to the green or blue-green group. Either partner in the association can live alone. Lichens live on rocks, walls, tree trunks, and other places where organic matter is scarce. The alga acts as an energy trap for sunlight, producing organic materials by photosynthesis.

(a)

FIGURE 13.1 Symbiosis between corals and the algal dinoflogellate *Zooxanthella.* (a) A common coral. (b) A cross section of coral tissue showing the cells full of zooxanthellae. epi, epidermis; m, mesaglea; en, endoderm; and zx, zooxanthellae. The zooxanthellae are confined to the endoderm. The mesaglea lies between the external epidermis and the internal endoderm. (From R. K. Trench, *Proc. Royal Soc. B,* 177:225 (1971). Both photos courtesy of R. K. Trench.)

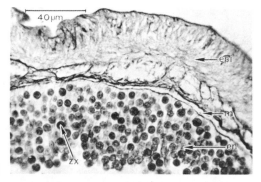

(b)

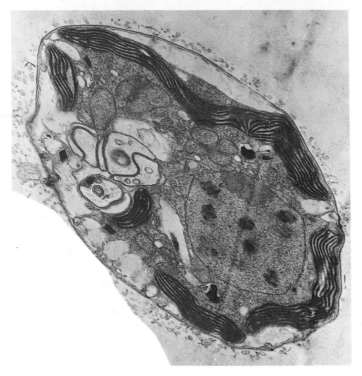

FIGURE 13.2 A cross section of *Zooxan-thella*. The alga has the structure of a typical dinoflagellate. Organic materials synthesized by photosynthesis are excreted to the coral cells. (Courtesy L. Loeblich.)

Corals provide another excellent example of symbiosis. The symbiont is a brown dinoflagellate alga, *Zooxanthella*, living within the coral polyps (Fig. 13.1). During the day the algae photosynthesize and produce soluble food that is transferred directly to the coral. At night the coral uses its tentacles to collect plankton. It is presumed that the coral provides the alga with essential growth factors. Figure 13.2 is an electron micrograph of the alga *Zooxanthella*.

Legumes are important agricultural plants that form symbiotic associations with the bacterium *Rhizobium* (p. 179).

THE RHIZOSPHERE AND PHYCOSPHERE

The Rhizosphere The root zone of all plants provides a nutrient-rich zone that supports a large and active microflora. This area of soil surrounding the roots is called the *rhizosphere*. The continuous natural attrition of plant root cells provides a flow of organic material into the soil. In addition, actively metabolizing root cells excrete large quantities of organic compounds. These materials are used as substrates by the rhizosphere microflora. The rich microbial population that grows along the plant roots is seen in Fig. 13.3.

The extent of microbial stimulation in the root zone is readily

234

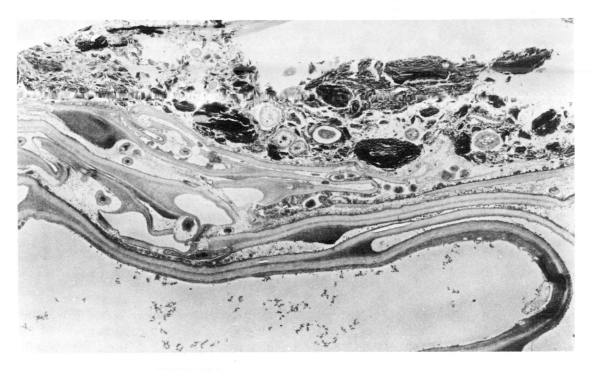

FIGURE 13.3 An electron micrograph showing the rhizosphere of a 16-week-old wheat root. The bottom of the photograph shows intact root cells. The center is a region of moribund distorted host cells. The top of the micrograph shows that these cells are inundated with a rich population of rhizosphere bacteria. (Courtesy A. Rovira.)

apparent by observation of the rhizosphere:soil (R:S) ratios of microorganisms. The rhizosphere population of young plants is small and R:S ratios vary from 5 to 15. In mature plants where root cell autolysis and excretion rates are high, however, the rhizosphere population is well developed and these plants have R:S ratios of 50 to 100.

The rhizosphere population has the important function of recycling nutrients for nutrition of the plant. Organic materials are mineralized or transformed by the microflora to yield essential growth factors for the plant. The production of CO_2 during respiration of the rhizosphere microorganisms plays an important role in plant growth.

The Phycosphere A microbial population analogous to the rhizosphere community exists in aquatic ecosystems. These microorganisms live on lysed cells and excretion products of algae. Those groups of organisms that live in close proximity to the algal

235

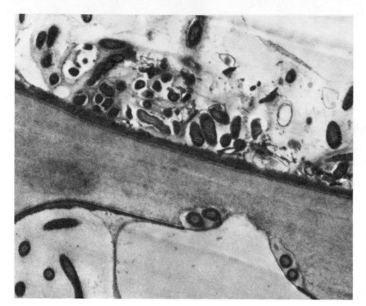

FIGURE 13.4   The phycosphere. A thin section of the benthic alga *Ulva* under the electron microscope shows a rich bacterial population growing on the surface. (Courtesy T. Waite.)

surfaces might be called the *phycosphere* microflora. This population is as important in the recycling of nutrients essential for algal growth as the rhizosphere population is for plant growth. The alga does not begin to excrete a sufficient quantity of organic material to support a microbial population until it is quite old. Older cells become leaky and excrete intracellular material.

Multicellular algae are composed of layers of cells and at any time some cells are dying and undergoing autolysis without serious damage to the algal sheet. They differ from unicellular algae in providing the microflora with both excretion and autolytic products. In this respect the multicellular algae are similar to plant roots. They have a phycosphere population throughout their lifetime and probably continually utilize nutrients mineralized by their phycosphere microflora. The extent that a multicellular alga is covered with microorganisms is seen in Fig. 13.4. This is a scanning electron micrograph of the benthic alga *Ulva*, which grows in abundance in eutrophic waters. The whole sheet of the alga is coated with bacteria.

ANTAGONISTIC INTERACTIONS

Under normal environmental conditions, microbial communities expand to the limits of space and nutrient supplies of the ecosystem. Populations may be as high as 10^7 per gram of rich soil and as low as 10 per liter of deep ocean water. In either case the population is as high as the system can support.

These communities are very diverse regardless of the population density. Diversity is maintained by three processes:

1. competition for space and nutrients,
2. production of toxic chemicals by one microorganism that kills another group of microorganisms,
3. intermicrobial predation.

Competition   Competition for nutrients is important in the determination of the dominant microorganism in a system. A wide range of saccharolytic bacteria might theoretically proliferate in a river containing molasses in the effluent from a fermentation plant. Only a small number of bacterial species in fact proliferates, however. Those bacteria that have the ability to assimilate the sugars in the molasses rapidly compete most successfully and predominate. In fully mixed waters one or two species would develop to the detriment of all others.

In reality, natural waters are never fully mixed. There is broad spatial heterogeneity that allows different species to proliferate while protected from their more aggressive cousins. The stalked bacteria survive well in the presence of far more rapidly growing bacteria by utilizing space more efficiently. These bacteria, which are sessile, grow by attaching to surfaces. They include *Metallogenium*, *Hyphomicrobium* and *Caulobacter* (Fig. 13.5). These bacteria act like sponges, slowly absorbing nutrients into the stalks as they flow by. They compete well on surfaces and are common in low nutrient waters.

FIGURE 13.5   The stalked bacterium *Caulobacter* growing on an electron microscope grid immersed in a pond. A flagellated motile cell has settled on the surface and is producing a stalk. Magnification 17,000X. (Courtesy P. Hirsch.)

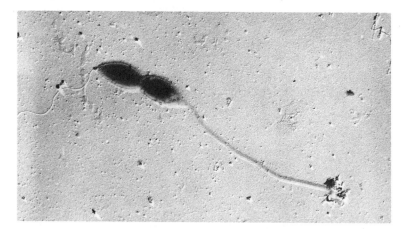

Antibiotics   Many microorganisms found in soil and fresh waters, as well as a high percentage of marine microorganisms, are capable of excreting organic materials that are antagonistic to other microorganisms. These chemicals may be either bacteristatic, inhibiting growth, or bactericidal, actively killing the other microorganism. These materials are called *antibiotics*. Penicillin, produced by the fungus *Penicillium*, is an antibiotic.

It would seem to be reasonable to assume that antibiosis is an important controlling process in microbial ecosystems. However, very low concentrations are excreted under natural conditions. The industrial production of antibiotics requires the use of rich complex media. When known antibiotic-producing bacteria are placed in soil or water, no significant quantities of antibiotics are detected.

The importance of antibiotic production as a general controlling process in nature is in doubt. Under specific environmental conditions, however, antibiosis can be significant. The actinomycetes are the most successful competitors for chitin as a substrate. Large numbers of actinomycetes are found on the chitinaceous shells of dead crustaceans in the sea. Under these circumstances the actinomycetes produce copious amounts of antibiotics, which presumably help to maintain the almost pure cultures of actinomycetes on the crustaceans. This population is transient and is soon replaced by another when the chitin is degraded.

Predation on Bacteria   The ability of microorganisms to prey on each other can be described as *intermicrobial predation*. This

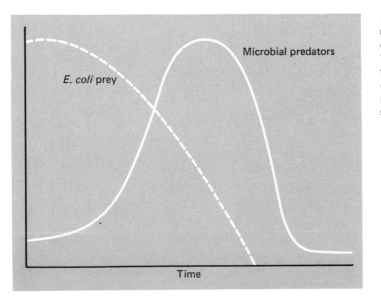

FIGURE 13.6   The response of native microbial predators to the introduction of a foreign microorganism *E. coli*. The *E. coli* cannot compete for nutrients and becomes a prey. The predators multiply at the expense of the prey and decline when the prey substrate is exhausted.

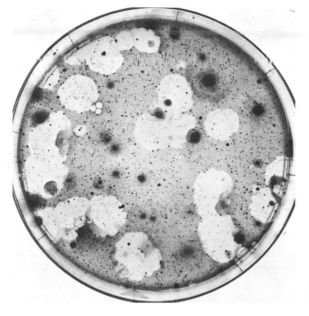

FIGURE 13.7 A petri dish of microbial predators grazing on a lawn of *Escherichia coli*. The clear zones are areas where the *E. coli* prey has been consumed.

process serves to weed out foreign microorganisms that are incapable of competing for nutrients. The most important example of this phenomenon is the destruction of enteric bacteria by native predators in the sea. Figure 13.6 illustrates how this population responds to the introduction of *Escherichia coli*. One might envisage a new sewage outfall disposing raw sewage to unpolluted coastal waters. Under natural conditions the predator population for *E. coli* in seawater is insignificantly small. When the coliform bacteria come in contact with the native microflora, there is a lag phase and then an exponential development of predators. These rapidly consume the enteric bacteria (Fig. 13.7). The predator population density will remain high if the sewage input continues. In the absence of a continuous supply of prey, Fig. 13.6 shows that the predator population returns to its original level. An identical situation exists in fresh waters and in soil. In all these environments three groups of predator microorganisms are involved:

1. Protozoa, particularly dinoflagellates and amoebae. These microorganisms are voracious and probably account for about 50% of the predatory activity.

2. Bacteria belonging to the genus *Bdellovibrio*. These unusual bacteria are tiny, highly motile, and obligately parasitic on other bacteria. They attach to the surface of the host bacterium, lyse the cell walls, and consume the cytoplasmic

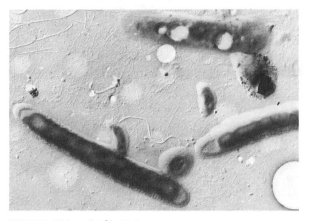

FIGURE 13.8 *Bdellovibrio,* a bacterial predator, attacking a lake bacterium. Note the flagellum of the predator with a characteristic basal long and terminal short wavelength. Magnification 25,000X. (Courtesy P. Hirsch.)

contents. These bacteria probably account for 30% of the predatory activity. A photograph of Bdellovibrios attacking a bacterium is shown in Fig. 13.8.

3. The least active group of predators on enteric bacteria is a group of bacteria that lyse the prey by enzymatic digestion of the cell walls. These are usually belonging to the genus *Pseudomonas* and may account for 20% of the predatory activity on coliforms.

Predation on Algae Native microorganisms that are poor competitors for nutrition are prime candidates for predation. Algae represent the most important example of this type of prey. Figure 13.9 provides representative curves of the relationship between algal biomass and predators in a lake.

Throughout the rising part of the algal growth curve there is predation on noncompetitive cells in the algal population. At the top of the curve an essential nutrient or light becomes limiting, and maximum biomass is achieved. At this point photosynthesis and predation rate are equal. In the declining phase predation exceeds photosynthesis. This phase occurs in natural waters in late summer. Algal cells begin to leak and form a large phycosphere population. Predators in this population attack the weakened cells and lyse them. The most common predators on algae are bacteria, that enzymatically degrade the cell walls. Thus, at the end of the growing season, a specific population of predators lyses the old algal population and releases the nutrients to the water column where they remain ready to fertilize the next year's algal growth.

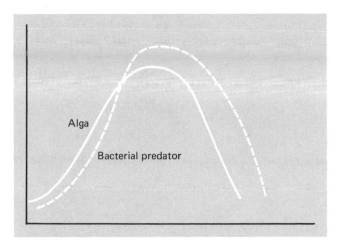

FIGURE 13.9 The relationship between algae and their bacterial predators. Noncompetitive cells are continuously eaten by the predators. At the peak of the algal growth curve the predators predominate and decimate the algal population. The predators overeat their prey and starve back to their normal low level.

Predation on Viruses It is probable, although unproved, that virus particles are susceptible to predation. The rate of attrition of viruses in natural waters is proportional to the density of the indigenous microbial population and is dependent on temperature. At temperatures below 10°C, when biological activity is low, viruses survive well in mixed populations, whereas at 25°C they are rapidly destroyed. One can only assume that they act as substrates for microbial predators in the water since they remain quite viable in sterile water.

CHEMICAL SIGNALS   Behavioral responses in animal communities are frequently controlled by chemical signals between animals. Chemicals controlling behavior within a species are called *pheromones*. Examples of pheromonal activity in aquatic animals include sex attractants produced by female lobsters, the pheromone produced by barnacles that causes other barnacles to aggregate in the same zone, and the alarm pheromone of mud snails that signals danger to other snails and allows them to escape (Table 13.1). The danger is usually the rising tide, which the snails escape by burying themselves. Signals between species are called *allemones*. This type of communication is seen in the search for food and particularly the detection of prey by predators. Prey usually excrete specific chemicals that act as attractants for their predator. Oysters produce an allemone that attracts its starfish predator.

TABLE 13.1 Some common pheromones.

Pheromones are chemicals that control behavioral responses between animals within a species. Pheromones control sexual, aggregation, alarm, and many other responses.[a]

Taxa	Activity of Pheromone	Chemical Nature of Pheromone
Protista *Volvox* sp.	Female substance induces gonidia to develop into sperm packets	High molecular weight, over 200,000; probably a protein
Paramecium bursaria	Mate recognition, by cilial contact	Apparently a protein
Aschelminthes *Brachionus* spp. (rotifer)	Recognition of females by males, followed by breeding	Not a protein; otherwise unknown
Annelida *Lumbricus terrestris* (earthworm)	Alarm and evasion; secreted in mucus	Unknown
Mollusca *Helisoma* spp. and some other aquatic snails	Alarm: self-burying or escape from water	Polypeptides from tissue; molecular weight about 10,000
Arthropoda Decapoda *Portunus sanguinolentus* (crab)	Sex attractant from female urine	Unknown
Cirripedia *Balanus balanoides* and *Elminius modestus* (barnacles)	Aggregation and settlement of larvae, by contact with pheromone on substratum	Protein
Arachnida Lycosidae (wolf spiders)	Female sex attractant	Unknown
Salticidae (jumping spiders)	Female sex attractant	Unknown

[a]From E. O. Wilson, *Chemical Ecology,* by E. Sondheimer and J. B. Simeone. Academic Press, N. Y., 1970.

242

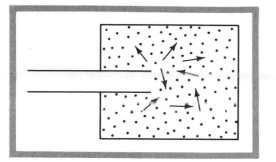 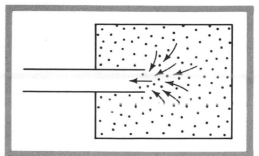

(a) Bacteria move randomly when no attractant is present.

(b) Bacteria orient and move along gradient of chemical diffusing from capillary.

FIGURE 13.10 A diagram showing bacterial chemotaxis. The motile bacteria orient and move toward the organic attractant. The bacteria are suspended in buffer on a microscope slide and the attractant is placed in a capillary. Within minutes after the capillary comes in contact with the bacteria, they begin to congregate around the attractant.

Bacterial Chemotaxis Chemical communication exists in the microbial world. The most primitive response in the attraction or *chemotaxis* of a motile bacterium to a chemical. All motile bacteria are attracted to one or more organic compounds. They may not necessarily metabolize the attractant but simply use it to find other substrates associated with it. Figure 13.10 illustrates a simple laboratory demonstration of bacterial chemotaxis. As soon as a concentration gradient of the attractant is established, the bacteria orient to it and swim toward the highest concentration. These concentration gradients are produced in decaying organic material. Degradation products attract an ever-increasing population of microorganisms. Physiologically deficient algal cells provide an excellent example. As we discussed on p. 240, these cells become leaky and attract a large population of bacteria.

Chemotaxis and Predation Microbial prey typically excrete chemical signals that specifically attract their prey. This phenomenon has been observed in the interaction between the diatom *Skeletonema* and its bacterial predator and between the fungus *Pythium* and its bacterial predator. Apparently autolytic products of the cell wall and leaked intracellular material act as specific attractants.

Chemotaxis and Sexual Behavior Although the extreme complexity of behavioral response seen in animal communities is not found among microorganisms, some complex chemical communication occurs at least among the eucaryotic protista (Table 13.1). The green alga *Volvox* produces a high molecular weight material that induces gondia to develop into sperm packets. The protozoan *Paramecium* produces a protein that permits mate recognition. It seems likely that similar signals exert control over behavioral activities in microbial communities, such as sexual interaction and aggregation.

SPECIES DIVERSITY

Under normal conditions the biological community is adapted to its environment. Complex integrated communities with a high species diversity are characteristic of healthy ecosystems. Microbial communities are extraordinarily diverse and versatile. Both benevolent and antagonistic interactions between the organisms maintain a high level of *biochemical* diversity. Most ecosystems contain many thousands of different microorganisms, each capable of some specialized biochemical function. Microorganisms may belong to the same species and yet differ in a biochemical function essential to the ecosystem. One strain of *Bacillus cereus* may be enzymatically capable of degrading the pesticide 2,4-D, while another strain acts on the pesticide 2,4,5-T.

Species diversity can be calculated even for bacteria. The most common index of diversity is the Shannon index $\overline{H}$:

$$\overline{H} = \Sigma \left(\frac{n_i}{N}\right) \log \left(\frac{n_i}{N}\right) - \Sigma \ P_i \ \log \ P_i$$

n_i = importance value for each species, measured as number of individuals, biomass, etc.

N = total of importance values for all species

P_i = n_i/N = importance probability for each species

SURFACES

Most natural habitats are nutrient deficient. This is true of oligotrophic lakes, rivers, the open ocean, and even soils. Under these circumstances it is not surprising to find that microorganisms congregate on surfaces. Concentrations of nutrients are found on clay particles and on organic debris. These particles serve as major zones of microbial activity. Indeed, if the particulate matter is filtered from natural waters, approximately 99% of the bacterial population is also removed.

The importance of surfaces for bacterial growth in seawater can be seen in Fig. 13.11. Different volumes of water are placed in

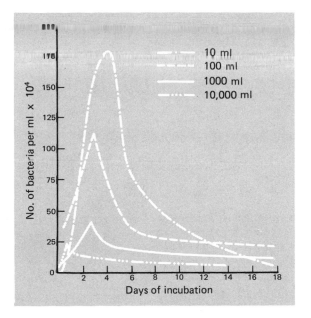

FIGURE 13.11 The growth response of marine bacteria in different volumes of seawater contained in similar containers. The growth and survival of the bacteria is directly related to the surface:volume ratio. (From C. ZoBell and D. Q. Anderson, *Biological Bulletin,* 71:324 (1936).)

glass containers of the same size so as to vary the surface:volume ratio. Bacterial activity is much greater at higher surface:volume ratios, indicating the importance of surfaces.

The mechanism of attachment of microorganisms to surfaces is not well understood. However, there appears to be three stages in the formation of surface films. The primary process is the attachment of a film of organic matter to the surface. All surfaces in contact with water immediately acquire this layer. The second stage is the attachment of a primary bacterial film. The normal bacterial population attaches temporarily to this surface by physicochemical forces. This attachment is usually by one end of the cell so that the bacterium is at right angles to the surface (Fig. 13.12). Motile bacteria may be attracted to the surface by chemotactic processes. Permanent attachment occurs when the attached bacteria exude extracellular polymers that bridge them to the surface (Fig. 13.13). This second stage of attachment occurs in a few hours. The third step involves the growth of slower growing microorganisms including stalked bacteria, overlying and replacing

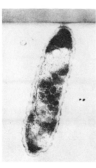

FIGURE 13.12 Initial attachment of a bacterium to a surface. The bacterium is held at one pole until polymeric bridges are formed. (Courtesy K. C. Marshall.)

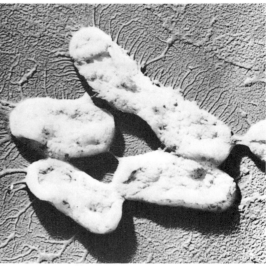

FIGURE 13.13 The second stage of attachment of bacteria to surfaces by polymeric fibrils. (Courtesy K. C. Marshall.)

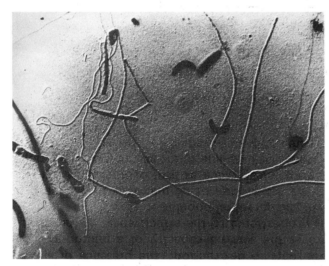

FIGURE 13.14 The third step in the formation of a surface film of microorganisms. The surface is coated with slower growing stalked bacteria. (From L. Young, Ph.D. Thesis, Harvard University, Cambridge, Mass., 1972.)

FIGURE 13.15 A scanning electron micrograph showing a late stage in microbial film formation. The surface is completely coated with a mixed population of microorganisms. (From L. Young, Ph.D. Thesis, Harvard University, Cambridge, Mass., 1972.)

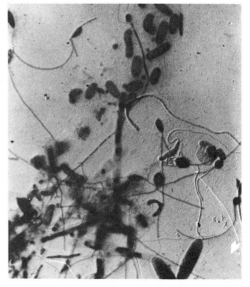

the primary film on the surface (Fig. 13.14). This growth usually takes as long as 7 days and culminates in the complete coating of the surface with stalked bacteria (Fig. 13.15). Ultimately the stalked bacteria are grazed by protozoa.

SUMMARY

1. Benevolent interactions enable microorganisms to nourish each other. These interactions lead to the development of diverse microbial communities. Benevolent interactions include commensalism, protocooperation, and symbiosis.

2. Antagonistic interactions prevent the dominance of any single microbial group. Antagonistic processes include competition, predation, and the production of toxic chemicals.

3. Chemical signals occur in the microbial world. All motile bacteria have chemoreceptors. Chemoreception is involved in microbial predator-prey relationships.

4. Microbial ecosystems display a high level of biochemical diversity. Many thousands of different organisms may be present in a soil or water habitat.

5. Surfaces are important for microbial growth. There is a direct relationship between surface:volume ratio and microbial activity. Development of surface films is a three-stage process.

FURTHER READING

M. Alexander, *Microbial Ecology*. John Wiley & Sons, Inc., New York, N. Y., 1971.

T. Brock, *Principles of Microbial Ecology*. Prentice-Hall Inc., Englewood Cliffs, N. J., 1966.

C. J. Krebs, *Ecology*. Harper & Row, Pub., New York, N. Y., 1972. See sections on community and population ecology.

STRESS
ON THE
MICROBIAL
COMMUNITY

The microbial community in natural ecosystems is brought under severe pressure by the products of human activities. We have discussed the biodegradation and biomagnification of many of these materials. At high concentration, chemical pollutants entering soil or water can be toxic to some components of the community and not to others. A similar effect is apparent when heated effluents enter water and form thermal gradients. These stresses alter the biological equilibrium.

EFFECT ON SPECIES DIVERSITY

In any ecosystem there is great diversity. Large numbers of species are present in the absence of stress. As the system becomes more perturbed, less adaptable species disappear and more resilient species take over the niche. In unperturbed systems the number of species or species diversity is high and the number of individuals in each species is relatively low. The inverse is true in highly perturbed systems.

The species diversity index gives a summation of the degree of stress on an ecosystem. The value of $\overline{H}$ for benthic organisms living in an unpolluted and polluted stretch of river is shown in Fig. 14.1. Close to the sewage outfall, $\overline{H}$ declines from the ambient of 4.0 to 1.0, indicating a sharp drop in species diversity. The diversity begins to increase 10 miles downstream and returns to normal 50 miles downstream from the outfall.

The fish population is the most sensitive and the algae and protozoa are the least sensitive to stress (Fig. 14.2). The diversity measurement of fish populations gives us information about slightly polluted waters. Algae or protozoa can be used in more heavily contaminated conditions. It must be emphasized that the

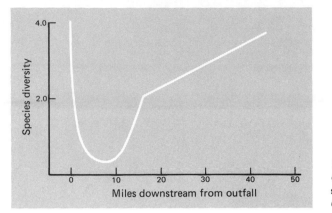

FIGURE 14.1 Species diversity index of benthic organisms as a measure of the damage caused by a sewage outfall. (From J. C. Wilhm, *J. Water Pollut. Contr. Fed.,* 39:1673 (1967).)

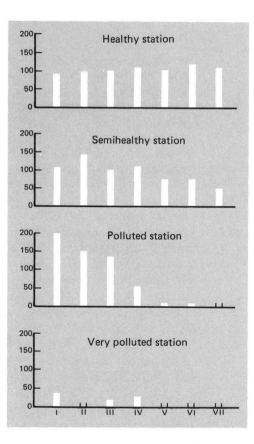

FIGURE 14.2 Population structure of aquatic communities under different degrees of pollution. Column I: Species of diatoms, blue-green algae, and green algae that are known to be tolerant of pollution. 4 species = 100% (these figures represent the average obtained from sampling several hundred areas in eastern coastal plain streams). Column II: Oligochaetes, leeches, and pulmonate snails. 6 species = 100%. Column III: Protozoa. 41 species = 100%. Column IV: Diatoms, red algae, green algae other than those included in Column I. 81 species = 100%. Column V: Prosobranch snails, triclad worms, and a few smaller groups. 11 species = 100%. Column VI: Crustaceans and insects. 47 species = 100%. Column VII: Fish. 20 species = 100%. (From J. Cairns, *Water Pollution Microbiology,* ed. R. Mitchell. John Wiley & Sons, Inc., N. Y., 1972.)

diversity index only tells us how much the system hurts. It does not tell us the nature or the permanence of the injury.

INORGANIC NUTRIENTS

The effect of nutrients on algal productivity was discussed in Chapter 11. Eutrophication causes major changes in the microbial population. The delicate balance between autotrophy and heterotrophy is upset by excessive productivity. Under normal conditions the productivity rate (P) is equal to the respiration rate (R). When large quantities of nutrients enter an aquatic ecosystem, a sequence of microbiological events is set in motion; this sequence is shown diagrammatically in Fig. 14.3.

1. P becomes greater than R. Algal productivity increases dramatically.

2. Simultaneously the species diversity index declines. The heterotrophic population has been drastically reduced. The best competitors among the algae for nutrients predominate in the algal bloom, lowering the species diversity of algae.

3. The supply of free nutrients in the water column is depleted.

4. Ultimately nutrients become limiting and maximal biomass is attained.

5. The overextended algal population crashes with a simul-

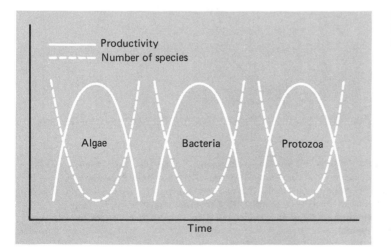

FIGURE 14.3 The sequence of microbiological events in an aquatic ecosystem with a large nutrient input.

Productivity
Number of species

Algae   Bacteria   Protozoa

Time

taneous massive rise in heterotrophic activity of bacteria and protozoa, and oxygen utilization. Eventually oxygen becomes depleted.

6. Nutrients are released to the water column and become available for the next cycle of algal productivity.

7. Species diversity remains low because of the intense heterotrophic activity with its associated strong selective pressure.

Nutrient pressure imposes extreme instability on the microbial community (Fig. 14.4). Soon after a large algal population develops and P becomes larger than R, it fails, giving way to an equally unstable heterotrophic community where R > P. This in turn fails, to be replaced again by P > R. A continuing input of nutrients to this system increases the size of the perturbation of P > R and R > P. Ultimately the whole ecosystem collapses.

The aquatic ecosystem returns to a P = R relationship coupled with low productivity and high diversity only when excessive nutrients are removed. This can be achieved by preventing further nutrient enrichment and waiting for natural biological and

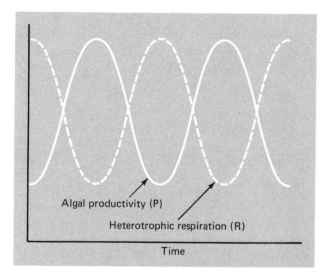

FIGURE 14.4 Instability imposed on the aquatic ecosystem by nutrient enrichment. Nutrients stimulate algal productivity so that P becomes greater than R. When the algae die, the released organic matter stimulates the heterotrophic microflora so that R becomes greater than P. The oscillation continues indefinitely. The stable situation P = R never occurs in polluted waters.

chemical processes to deprive the water column of an essential nutrient. Nutrient deprivation can be accelerated by treatment processes that are discussed in Chapter 16.

ORGANIC SUBSTRATES

The addition of sewage or other easily biodegraded organic compounds to natural waters alters the P:R balance in favor of respiration. The heterotrophic population predominates causing a decline in algal productivity. Figure 14.4 illustrates the sequence of events. The algal biomass begins to decline soon after the addition of organic matter, to be replaced by an equivalent population of heterotrophic microorganisms. The heterotrophs decline when all the organic substrate has been utilized. Since the inorganic products are not removed, however, the system does not return to P = R. The system has been enriched with algal nutrients and the consequence is increased by algal productivity and P > R. The organic matter has been removed but the ecosystem is now out of equilibrium. It will continue to oscillate between P > R and R > P until the inorganic nutrients are removed.

There is a close relationship between species diversity of the heterotrophic and autotrophic populations and concentration of organic substrate. As the size of the bacterial population increases in response to organic substrate load, the number of species declines (Fig. 14.5). The species diversity returns to normal when

FIGURE 14.5 The inverse relationship between the increased bacterial population developed in response to an organic matter load and the number of species. The species diversity declines under the substrate pressure and returns to normal when the substrate concentration declines.

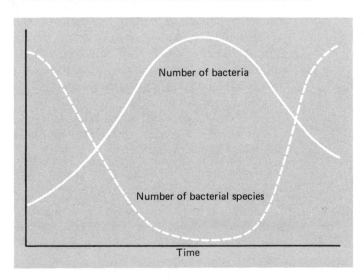

FIGURE 14.6 Predation of sea urchins on the giant kelp *Macrocystis* in sewage enriched waters off the coast of southern California. (Courtesy W. J. North.)

most of the organic substrate has been degraded. Conversely, when the productivity of the algae is depressed by the heterotrophic activity, the number of species increases. When the organic matter is removed, the algal species diversity declines below its normal level because of the increased inorganic nutrient concentration and the increased productivity.

The sensitivity of complex ecosystems to nontoxic organic pollutants is illustrated by the erosion of the giant kelp (*Macrocystis*) beds off the coast of southern California. Many square miles of the Pacific Ocean were covered by kelp, providing excellent fish breeding grounds, until sewage outfalls were built. The kelp died off rapidly in the vicinity of the sewage.

Destruction was caused by the proliferation of sea urchins (Fig. 14.6) that preyed on the kelp stipe, cutting it free from its holdfast and allowing large quantities of seaweed to float away. The decline of the kelp was followed by a reduction in numbers and species of fish.

The kelp recovers when the flow of sewage is stopped. In the presence of sewage, recovery can be achieved by liming the water to kill the sea urchins.

TOXIC CHEMICALS   Heavy metals, pesticides, and a large number of other inorganic and organic toxic industrial effluents are disposed into natural waters. The most drastic effect of these chemicals is to kill both the microflora and the higher organisms. Indirect effects include

1. concentration,
2. selective toxicity,
3. behavioral aberrations.

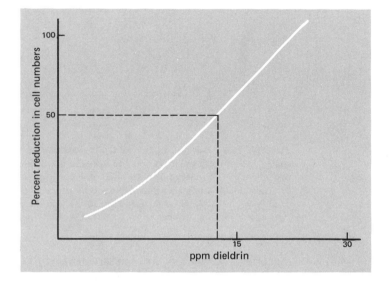

FIGURE 14.7 The toxic effect of the insecticide dieldrin on the diatom *Navicula.* The diatom survives very high concentrations of the pesticide. The population is only reduced by 50% with 12 ppm of dieldrin in the water. The ability of the surviving algae to absorb dieldrin would make them particularly dangerous to fish. (From J. Cairns, Jr., *Mosquito News,* 28:177 (1968).)

Concentration The insecticide dieldrin is commonly used for mosquito control. The toxicity of this chemical for the diatom *Navicula* is seen in Fig. 14.7. A concentration of 12 ppm dieldrin in the water kills only 50% of the diatoms. The algae absorb and concentrate the dieldrin, posing a threat to fish and other organisms that consume the diatoms.

Selective Toxicity Some chemical pollutants display a selective toxicity for certain algae. For example, DDT and PCB's kill the

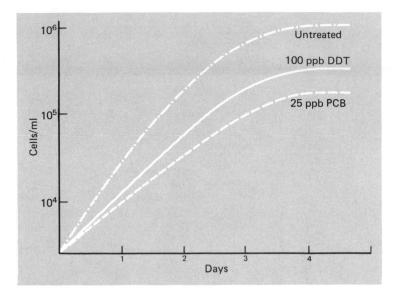

FIGURE 14.8 Selective toxicity of DDT and PCB's for the diatom *Thalassiosira* in mixed cultures. DDT inhibited the diatom at a concentration of 100 ppb. PCB's were inhibitory at 25 ppb. Another alga, *Dunaliella,* in the mixture was totally unaffected by these concentrations of DDT or PCB's. (From J. L. Mossler, et al., *Science,* 176:533 (1972). Copyright by The American Association for The Advancement of Science.)

diatoms *Thalassiora* at concentrations in parts per billion (Fig. 14.8). The green alga *Dunaliella* is resistant to low concentrations of these chemicals. In uncontaminated mixed cultures the diatom outgrew the green alga by a factor of 8 or 9. The ratio is changed to between 3 and 4 by 10 ppb of DDT. The same concentration of PCB changes the ratio to between 1 and 2.

These are quite common concentrations of DDT and PCB for marine waters. Any change in the dominant algae at the base of the food web will affect the nutrition of zooplankton and of herbivorous fish. The effect on the composition of the animal population could be profound.

Oil spills cause a 95% reduction in the benthic fauna in the region of the spill. Shellfish, crab worms, and bottom-living fish are all killed. An analysis of the polychaete population following the oil spill at West Falmouth, Mass., showed that most of the polychaetes were eradicated by the petroleum. One opportunistic species, *Capitella capitata,* survived the spill, however, and predominated until the oil had largely disappeared (Fig. 14.9). At that time the natural polychaete population returned and *Capitella*

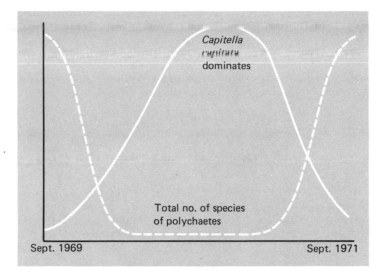

FIGURE 14.9 Destruction of the subtidal polychaete population by an oil spill in West Falmouth, Mass. The polychaete *Capitella capitata* survived the spill and temporarily dominated the empty niche until the effect of the spill had diminished. (From H. L. Sanders, Woods Hole Oceanographic Institution, Report No. 72-20, 1972.)

declined to its normal level. This provides another example of the maintenance of productivity in perturbed ecosystems. The form of the productivity, however, may seriously distort the ecological balance of the system.

Behavioral Aberrations   The effects of toxic pollutants on chemical signals are not well understood. Crude oil is known to interfere with the sexual responses of lobsters and with the detection of oyster prey by starfish.

The ability of motile bacteria to detect and move toward attractants has been discussed on p. 243. This process is blocked by many different chemical pollutants including crude oil, pesticides, and polychlorinated biphenyl compounds. Apparently pollutants competitively attach to chemoreceptors, blocking them so that they cannot detect their normal attractants. Chemotactic responses are inhibited by concentrations of chemicals in the water that are too low either to kill the bacteria or to affect motility adversely (Table 14.1). The ecological effects of this phenomenon could be great. Chemical signals play an important role in the control of animal behavior in natural waters. It is be-

TABLE 14.1 Inhibition of bacterial chemotaxis by pollutants in sea-water.

The attraction of the motile marine bacteria to organic matter is inhibited by extremely low concentrations of benzene. Heavy metals, the pesticide 2, 4-D, and crude oil only inhibit chemotaxis when they are present in the water at high concentrations. Portions of the crude oil are very insoluble so that it is difficult to determine the active concentration.

Pollutant	Concentration	Number of Bacteria Attracted to a Mixture of Organic Compounds
None	—	1,000,000
Benzene	10^{-5} M	500,000
Copper salts	10^{-2} M	80,000
Lead salts	10^{-2} M	30,000
2,4-D	10^{-3} M	80,000
Crude oil	0.2%	270,000

coming apparent that these signals are also utilized in the control of microbial communities. Large-scale interference with chemoreceptors would seriously upset the structure of both animal and microbial communities.

The effect of toxic chemicals on the microbial community is dependent on (1) concentration; (2) time of contact; (3) physical conditions, temperature, pH, O_2, etc.; and (4) the physiological condition of the organism. Interaction of these parameters and between chemicals frequently obscure the origin of disturbance of the community.

THERMAL PERTURBATIONS

The development of nuclear power plants has dramatically increased the quantity of heated effluents. Nuclear power production is much less efficient than fossil fuels and releases almost 60% more unusable heat than conventional plants. Diffusion of this low level energy requires large quantities of cooling water.

The effects of this form of disturbance can be estimated more easily than other perturbations. At slightly increased temperatures, no effect is noted. As the temperature is raised further, the marine fauna is affected. Some fish die; others have their reproductive processes impaired; while others may contract diseases that they resist in normal circumstances. Ultimately the temperature becomes so high that all fish die.

Thermal pollution in highly eutrophic waters or in waters enriched with sewage or other organic matter ultimately causes oxygen depletion. The increased temperature stimulates biological productivity and hence oxygen utilization from respiration of heterotrophic microorganisms. Even in oligotrophic lakes massive thermal changes provide the potential for destratification. Nutrients would be raised from the sediments to the euphotic zone, stimulating excessive algal productivity and leading to oxygen deficiency. Heat also lowers the solubility of oxygen so that less is available for the aquatic community.

The most serious effect of thermal pollution on the microflora is the shift in equilibrium of the microbial community. Figure 14.10 shows this effect dramatically. The predominant algae in cool fresh waters, the diatoms, begin to disappear as the temperature rises above 21°C. At 32°C, the diatoms have disappeared and are replaced by green algae. These in turn are replaced by blue-green algae at 37°C.

The development of blue-green algal blooms is undesirable because of the toxicity of many species and the unpleasant taste and odors produced by them.

FIGURE 14.10   The equilibrium of the photosynthetic microbiological community is moved by thermal pollution. The diatom is the dominant alga at low water temperatures. Between 25 and 30°C the green algae predominate. Above 37°C the blue-greens are the only algae growing in the water. (From J. Cairns, Jr., *Ind. Wastes,* 1:156 (1956).)

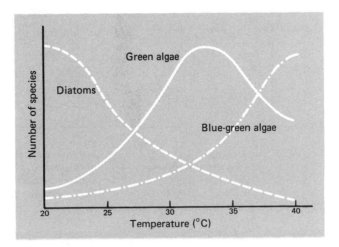

RADIOACTIVE ISOTOPES

The potential output of radioactive isotopes by atomic power plants was discussed in Chapter 12. The concentration of radioactive isotopes in the biosphere has a potential to alter the structure of an aquatic community. Radioactive wastes may accumulate in increasing amounts as more atomic power plants are built. Stable isotopes could concentrate in algae and bacteria in extremely low concentrations. The process of biomagnification may lead to tenfold or a hundredfold concentration in higher organisms.

DIVERSITY AND STABILITY

Throughout this chapter, I have emphasized the inverse relationship between stress on the ecosystem and species diversity. What is the connection between these two parameters and the stability of the ecosystem? Microbial communities with low stress and high diversity are very stable and unproductive. High productivity—low diversity ecosystems are unstable. This principle is

FIGURE 14.11 The relationship among productivity, diversity, and stability in aquatic ecosystems. Young highly productive communities contain few species and are quite unstable. Old communities are relatively unproductive, have many species, and are quite stable. Nutrient-rich waters represent artificially young communities. (From W. Stumm and E. Stumm-Zollinger, *Water Pollution Microbiology,* ed. R. Mitchell. John Wiley & Sons, N. Y., 1972.)

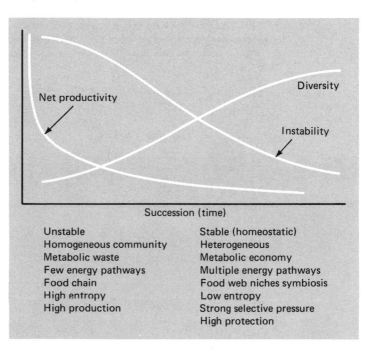

Unstable	Stable (homeostatic)
Homogeneous community	Heterogeneous
Metabolic waste	Metabolic economy
Few energy pathways	Multiple energy pathways
Food chain	Food web niches symbiosis
High entropy	Low entropy
High production	Strong selective pressure
	High protection

summarized in Fig. 14.11. Eutrophic waters and ecosystems contaminated with either organic substrates or toxic chemicals have few species.

The nutrient-rich habitats are also highly productive. However, toxic compounds reduce productivity.

It is tempting to generalize the relationship between diversity and stability. Microbial communities appear to follow the rule closely. The theory does not hold for higher organisms, however. Coral reefs produce extremely complex diverse ecosystems, but reefs are very unstable. Conversely, the *Spartina* salt marsh has a very low diversity and high productivity and is quite stable.

SUMMARY

1. The species diversity in a body of water is a good index of stress. Ecosystems polluted by toxic chemicals or by nutrient enrichment have a lower species diversity.

2. The productivity:respiration ratio is also used as an index of pollution. In polluted waters the P:R ratio is greater or less than unity.

3. Toxic chemicals and increased temperatures interfere with the ability of bacteria to respond to chemical signals. This interference may have important ecological significance.

4. Thermal pollution changes the structure of both the heterotrophic and photosynthetic microbial populations. Increased algal productivity and a change in the dominant algal population drastically alters the fish population.

5. Radioactive isotopes are biomagnified in the food chain. These alter the structure of the biological population.

6. There is a relationship between diversity and stability in microbial ecosystems. More diverse systems tend to be stable and unproductive.

FURTHER READING

J. Cairns, Jr. and G. Lanza, "Pollution Controlled Changes in Algal and Protozoan Communities," in *Water Pollution*

Microbiology, ed. R. Mitchell. John Wiley & Sons, Inc., New York, N. Y., 1972.

P. A. Krenkel and F. L. Panker, *Biological Aspects of Thermal Pollution*. Vanderbilt University, Nashville, Tenn., 1969.

R. Mitchell, "Ecological Control of Microbial Imbalances," in *Water Pollution Microbiology*, ed. R. Mitchell. John Wiley & Sons, Inc., New York, N. Y., 1972.

National Academy of Sciences, *Eutrophication: Causes, Consequences, Correctives*. Washington, D. C., 1969.

H. T. Odum, *Environment, Power and Society*. John Wiley & Sons, Inc., New York, N. Y., 1971.

CONVENTIONAL WASTE AND WATER TREATMENT

Biological treatment has been used as a means of sewage disposal for over a century. The conventional sewage treatment plant makes use of microorganisms to oxidize organic matter in feces to the mineral form and to entrap a portion of these minerals in bacterial cells. The carbon in these cells is microbiologically reduced to methane gas by anaerobic digestion.

Drinking water treatment serves to kill potential pathogens and to eradicate iron and manganese as well as organic compounds producing tastes and odors in the water.

SEWAGE TREATMENT   Sewage is composed of (1) human fecal material; (2) domestic wastes including food wastes and wash water; and (3) industrial wastes.

The industrial effluents include biodegradable, recalcitrant and toxic chemicals.

Biological treatment of sewage is used to (1) eradicate human pathogens carried in sewage and (2) oxidize the organic matter that may be present as microbial cells, insoluble debris, or soluble organic matter. The process is shown diagrammatically in Fig. 15.1.

There are five stages in biological waste treatment:

1. primary treatment,
2. biological oxidation,
3. flocculation and settling,
4. anaerobic digestion of flocculated cells,
5. chlorination of the effluent.

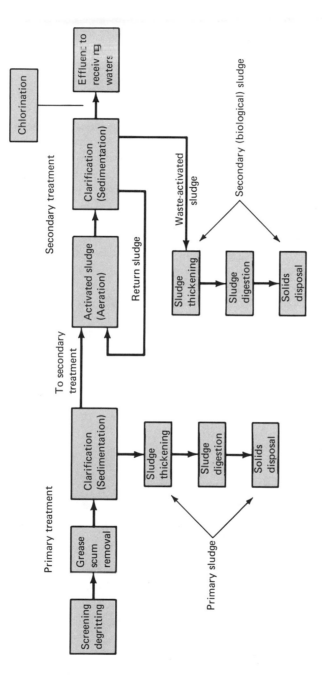

FIGURE 15.1 Biological waste treatment. Primary treatment consists of removal of grease and scum followed by sedimentation of large particles. Secondary treatment involves oxidation in the activated sludge process by microorganisms of the colloidal and dissolved organic matter remaining in the sewage. The microbial cells flocculate and sediment to produce sludge, which is digested anaerobically to yield methane gas. The secondary effluent may undergo advanced treatment (see Chapter 16) or be chlorinated and disposed to the receiving waters. (From "Cleaning our Environment—the Chemical Basis for Action," a report by The Subcommittee on Environmental Improvement, Committee on Chemistry and Public Affairs, American Chemical Society, 1969, p. 107. Reprinted by permission of the copyright owner.)

Primary Treatment   This is a physical process in which suspended solids are removed by passage of the sewage through screens. The sewage is placed in sedimentation tanks to allow settling of solids. Only colloidal organic matter remains after primary treatment. Frequently, primary settling and chlorination are the sole treatment sewage receives before marine disposal.

BIOLOGICAL OXIDATION

Activated Sludge   Secondary treatment of sewage involves the biological oxidation of a heterogeneous continuously changing mixture of organic compounds. The most common   process, *activated sludge*, depends on an enrichment culture of bacteria to oxidize the sewage.

The treatment tank consists of a highly aerated fully mixed system in which the effluent is in continual contact with a native population of microorganisms, the activated sludge. This population is maintained by returning a portion of the settled sludge to the tank.

The secondary treatment plant behaves as a chemostat.   The rate of mineralization of the organic substrate is dependent on the detention time. Bacterial enzyme kinetics serve as a guide to determine the optimum size and detention time for a specific organic matter load on the facility.

$$K = \frac{1}{s} \cdot \frac{ds}{dt} = K_{max} \frac{C}{C_k + C}$$

describes the relationship in its simplest terms.

K = the exponential growth rate of the microorganisms

C = concentration of limiting substrate at time t

K_{max} = the maximum growth rate when substrate is not limiting

C_k = Micheles-Menten constant, which is the concentration of limiting substrate at which K is $0.5\ K_{max}$

S = concentration of microorganisms

The complexities of enzyme repression interfere with this reaction, however. Enzyme repression is discussed on p. 144. Prediction of mineralization rates in constantly changing mixed substrates with complex bacterial populations must be at best tenuous.

The oxidation of organic substrates in the activated sludge tank is accompanied by the development of a large population of microorganisms taking part in the oxidation. An ecological succession occurs in the activated sludge plant. Figure 15.2 illustrates that protozoa predominate initially. They consume the microorganisms in the sewage. The protozoa decline and are replaced by a wide range of heterotrophic bacteria. These in turn are preyed on by other groups of protozoa. Finally the protozoa begin to die. Protozoan and bacterial polymers accumulate in the tank and cause the microbial cells to flocculate.

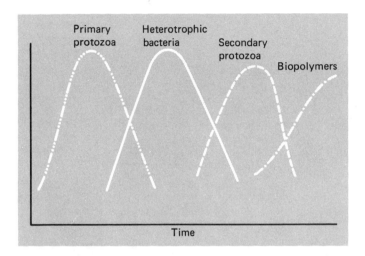

FIGURE 15.2 The ecological succession associated with activated sludge treatment of wastes. The large number of bacteria in the waste stimulates development of a protozoan population. Heterotrophic bacteria develop on the protozoan products and on dissolved organic matter. Finally, a secondary protozoan population forms before flocculation.

Flocculation   The rate of microbial cell multiplication and mineralization of organic matter depends on the carbon:nitrogen ratio described in Chapter 8. When the microorganisms become old and nitrogen becomes limiting in the tank, extracellular polymer excretion increases. The polymers are mainly in the form of bacterial polysaccharides. In addition, many old cells lyse and release polymeric nucleic acids and polypeptides. These polymers serve to flocculate the bacterial cells. The relationship between cell flocculation and polymer concentration is stoichiometric (Fig. 15.3). Flocculation depends on the concentration of both cells and polymers.

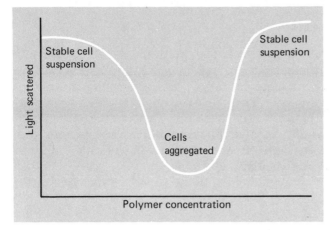

FIGURE 15.3 The effect of polymers on cell flocculation. There is a stoichiometric relationship between polymer concentration and degree of aggregation. Too little or too much polymer prevents flocculation.

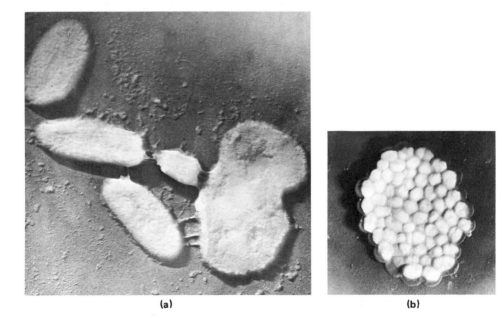

(a)   (b)

FIGURE 15.4 Polymeric bridging causes bacterial flocculation. (a) Bacterial extracellular polysaccharides. (b) A bacterial floc caused by bridging of polymers. (From R. Harris, Ph.D. Thesis, Harvard University, Cambridge, Mass., 1970.)

Figure 15.4 shows the bridging of bacterial cells by extracellular polymers. These bridges yield the closely packed flocs of cells. The flocs or aggregates lose the colloidal stability of their bacterial components and settle out of the liquid. The same result can be achieved using artificial polymers to bridge the bacterial cells. Artificial polymers are sometimes used to increase the efficiency of removal of bacteria in the activated sludge process.

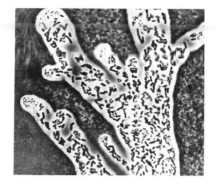

FIGURE 15.5 *Zooglea ramigera,* a common bacterium in activated sludge. The photograph shows the large amounts of extracellular polysaccharide produced by these bacteria, making them important in the flocculation process. (Courtesy R. F. Unz.)

Zooglea ramigera, a rod-shaped bacterium, is common in activated sludge and produces copious amounts of extracellular slime. A photomicrograph of *Zooglea* is shown in Fig. 15.5. It has been suggested that *Zooglea* is the most important organism in bacterial flocculation.

The flocculated cells are settled out for further treatment and the mineralized effluent (with 95% of the available organic matter removed) is taken out of the activated sludge tank, chlorinated to kill any remaining pathogens, and returned to the natural waters.

Trickling Filters   The trickling filter process illustrated in Fig. 15.6 offers an alternative to activated sludge treatment. These filters are made from a bed of gravel more than 4 feet thick over which the sewage slowly trickles. The sewage is sprayed onto the bed so as to give maximal aeration. The gravel becomes coated with a zoogleal film. The microbial population on the surface consists of a mixture of bacteria, fungi, and protozoa. These microorganisms are responsible for the oxidation of the organic matter in the sewage. Trickling filters are frequently used to treat cannery and other food industry wastes.

FIGURE 15.6   A trickling filter. The sewage trickles from a sprinkler onto a bed of sand covered with gravel. The high rate of aeration and the microbial film on the sand surface account for the efficiency of purification.

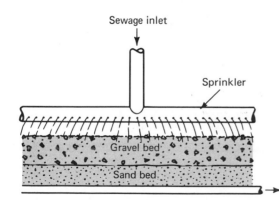

Sewage inlet

Sprinkler

Gravel bed

Sand bed

Effluent outlet

Oxidation Ponds Oxidation ponds are common in rural areas where there is no limitation on space. Both domestic sewage and industrial wastes are oxidized by treatment in shallow ponds.

The process can be seen in the diagram in Fig. 15.7. Heterotrophic bacteria convert the organic matter to cell material. The mineral products of organic matter decomposition support an algal population that replenishes the oxygen depleted by the heterotrophic bacteria. It is important not to make the depth of the pond greater than the euphotic zone so as to maintain aerobic conditions. Typical depths are less than 10 feet. The residence time in these ponds may be as long as a week. Effluents containing the oxidized products are removed at regular intervals so that the oxidation pond works as a batch culture. Algae and bacterial cells flocculate, as in an activated sludge plant, and settle to the bottom of the pond.

FIGURE 15.7 The biological processes in an oxidation pond. The photic zone is highly productive. Dead algae fall into an aerobic heterotrophic zone where they are partially degraded. Partially degraded algae and many dead bacteria fall to the anaerobic bottom of the pond.

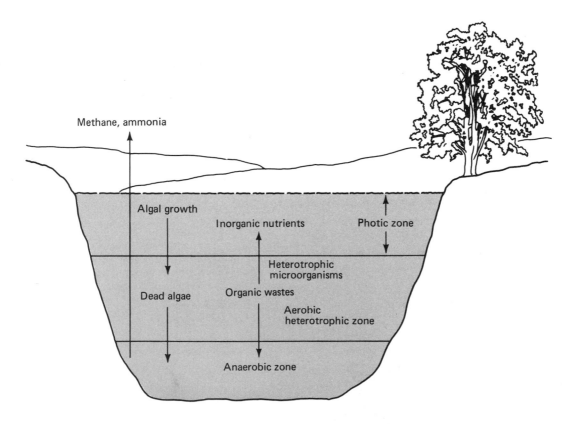

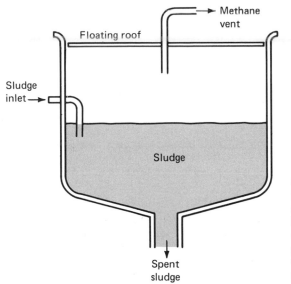

Methane vent

Floating roof

Sludge inlet →

Sludge

Spent sludge

FIGURE 15.8 An anaerobic sludge digester. The sludge is fed into the tank, which is held under completely anaerobic conditions. The methane produced by the methane bacteria is stored under a floating roof until it is removed from the top of the digester. The spent sludge, which is primarily composed of low nitrogen cell debris, is removed from the bottom of the tank. The tank is usually heated to maintain a thermophilic microflora.

ANAEROBIC DIGESTION

Sludge Digestion The microbial cells that settle out in primary and secondary activated sludge treatment are collected. This sludge is degraded by *anaerobic digestion*.

The anaerobic digestion tank shown in Fig. 15.8 is used for this process. Sludge is piped into the tank, which is maintained free of oxygen.

The proteins, fats, and carbohydrates that comprise the microbial cells in the sludge are degraded by a mixed flora of anaerobic bacteria to yield volatile organic acids, alcohols, and hydrogen sulfide. The genus *Clostridium* predominates. The methane bacteria convert these acids to methane gas (p. 147). Digestion occurs most rapidly at temperatures of 50-60°C, which favor thermophilic bacteria. The optimum pH for the process is 7.0 The methane is burned and the energy produced is frequently used to run the treatment facility.

Septic Tanks In rural areas, individual homes treat their sewage in septic tanks as seen in Fig. 15.9. These are simple settling tanks placed in the ground near the home. The sewage passes through the tank slowly so that solids are settled out within the tank. The effluent leaving the tank and percolating into the soil is at least partially free of suspended solids.

The sludge within the septic tank is maintained under anaerobic conditions. Anaerobic bacteria in the tank convert the sludge to organic acids and hydrogen sulfide. These products are dis-

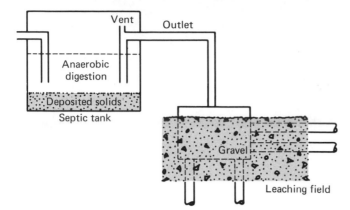

FIGURE 15.9 A septic tank, used for disposal of sewage in rural homes. The sewage is degraded by anaerobic digestion. The treated effluent is disposed into a leaching field of soils with good permeability. The effluent pipes are usually set in gravel.

tributed into the soil with the sewage effluent. Septic tanks do not necessarily destroy all pathogens. Drainage fields should not allow effluents from the septic tank to contaminate drinking water supplies.

Imhoff Tanks The Imhoff tank is a modification of the septic tank (Fig. 15.10). The tank is divided into two sections. The upper section acts as a settling tank. Digestion occurs in the lower section where conditions are more anaerobic. Gaseous products are

FIGURE 15.10 An Imhoff tank. The settling compartment is separated from the digestion compartment. Solids move down into the digestion compartment but are prevented from being carried back up into the settling compartment. Stabilized solids are removed from the bottom of the tank. Gases are vented from the top.

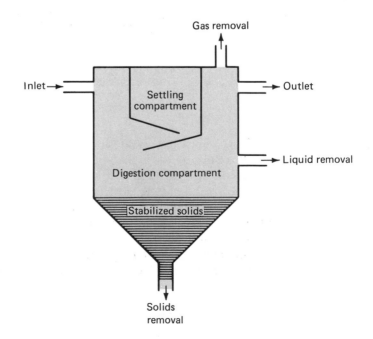

removed by a gas pipe. The Imhoff tank is quite efficient, allowing complete digestion in 2 to 4 hours. It requires expert maintenance, however, so it can only be used by municipalities. It has been largely replaced by secondary treatment followed by sludge digestion.

DISINFECTION

The final stage of waste treatment is disinfection. It is not unusual for a municipality to provide only primary treatment or no treatment at all before disposing of sewage. It is rare, however, for a municipality to allow either treated or untreated effluents to enter natural waters without disinfection to ensure destruction of pathogens. Effluents leaving the plant usually contain 0.5 mg/liter residual chlorine after 15 minutes of contact time. This level appears to be a safe average. Approximately 99.9% destruction of *Eschericia coli* is achieved by this level of disinfection.

DRINKING WATER TREATMENT

Our source of drinking water may be either surface water or ground water. In either case the potential for disease is always present. Fecal contamination of drinking water supplies is not uncommon. Our phenomenal success in controlling waterborne disease, despite the rapid rate of urbanization, is directly related to the treatment of our municipal drinking water supplies, which supply almost 80% of the population of the United States. In addition to microbial contamination, drinking water may contain organic or inorganic colloids. Clays, microbial debris, and reduced iron and manganese compounds are commonly found in drinking water supplies.

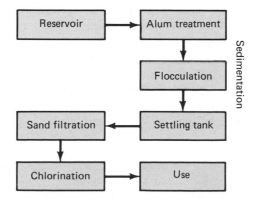

FIGURE 15.11 Flow diagram showing the treatment of drinking water. Alum is used to flocculate iron, manganese, and bacterial cells. They are sedimented in settling tanks. Fine colloids are removed by sand filtration. The water is chlorinated before use.

Treatment involves three stages:

1. sedimentation,
2. filtration,
3. chlorination.

The process is shown in Fig. 15.11.

Sedimentation   Reduced iron and manganese are frequently present in waters containing high concentrations of organic matter. They impart a brown coloration. Water containing iron and manganese salts also has an unpleasant taste. Permanganate is added as an oxidant together with a chemical to neutralize the pH. At neutral pH, reduced iron and manganese compounds are rapidly oxidized.

Flocculation may be achieved by use of alum, ferrous sulfate, ferric chloride, or synthetic polyelectrolytes. Bacteria, organic colloids, and oxidized iron and manganese all coagulate in the presence of flocculents. The flocculent neutralizes the negative charge on the colloid surface. The colloids form a bridge to the flocculent and become destabilized. The flocculated material is settled out in sedimentation ponds.

Filtration   The flocculated colloids remaining in the water are removed by filtration through fine-grained sand. The sand grains accumulate a microbial layer similar to that found in trickling filters. This acts as a fine filter to hold the floc. The filter is normally composed of an upper layer of about 12 in. of gravel with a bottom layer 1 to 2 feet thick of fine sand. A combination of a layer of sand above a layer of anthracite is sometimes used. The filter is backwashed regularly to remove adsorbed floc. Backwashing usually is carried out at least once a day.

Additional treatment is required if excessive organic taste and odor compounds are present. These may result from growth of actinomycetes or algae. Passage through a column of activated carbon removes these contaminants.

Drinking water must not have excessive nitrate present because of the danger of methemoglobinemia. The concentration of nitrate, particularly in rural areas, should be determined routinely.

Disinfection   All municipal drinking water supplies must be disinfected. This precaution is taken even when no obvious evidence of fecal contamination is present. The redundancy built into the system ensures that the public is never exposed to human pathogens through their drinking water supply.

GROUNDWATER RECHARGE

In many cities untreated wastes have been disposed into surface waters for centuries. In some arid zones alternatives to conventional sewage treatment that yield recycled water have been found. This treatment method, *groundwater recharge*, depends on soil infiltration to partially purify the water. The purified water ultimately returns by gravity flow to the subsurface aquifer and is mixed with natural groundwater (Fig. 15.12).

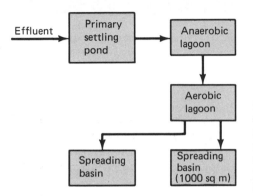

FIGURE 15.12 A flow diagram showing typical water purification by groundwater recharge. The sewage is treated before infiltration in a succession of ponds involving primary settling anaerobic decomposition, and aerobic oxidation.

The Process Initially the sewage is treated in primary settling ponds. This is followed by organic matter oxidation in a series of two oxidation ponds. The pond effluent has a large population density of algae.

The next phase of treatment involves infiltration of this material through natural sand and soil columns into the aquifer. This is achieved in large infiltration ponds (Fig. 15.13).

Infiltration removes the bulk of the algae, dissolved organic matter, and enteric bacteria and viruses in the effluent, by a combination of adsorption onto sand and soil particles and oxidation by a sessile microflora on the particles.

The water recovered from the aquifer is low in organic matter. There is no appreciable count of enteric bacteria. It is possible that some enteric viruses survive infiltration. These can be eradicated by either long-term detention or chlorination of the groundwater. The oxidation of large quantities of organic matter

FIGURE 15.13 Infiltration of partially treated sewage in spreading basins during groundwater recharge operation. (Courtesy Los Angeles Sanitation District.)

yields high concentrations of salts, particularly nitrates. When this water is to be used for drinking purposes, it is mixed with very soft water to yield drinking water with nitrate and other salt concentrations acceptable by public health standards.

A possible relationship exists between water hardness and arteriosclerosis. Groundwater recharge causes a continual increase in salt concentration of the water. If water hardness proves to be a cause of heart disease, then recharged water will require more extensive treatment if it is to be used for drinking purposes.

Clogging of Soil   Infiltration ponds rapidly clog near the soil-water interface. Clogging is accompanied by the accumulation of a black ferrous sulfide precipitate in the clogged layer. The ferrous sulfide is indicative of reducing conditions in that zone. Clogging is caused, however, by bacterial polysaccharides. The sequence of biological events leading to clogging is shown below.

1. Production of large quantities of bacterial polysaccharides.
2. Depletion of oxygen in the zone of highly active metabolic activity.
3. Inhibition of aerobic polysaccharide-decomposing bacteria.
4. Accumulation of polysaccharides causing clogging.

5. Reduction of ferric iron and sulfate by microorganisms to ferrous sulfide in the anaerobic zone.

When the system is maintained in the aerobic phase, no clogging occurs because the polysaccharide produced is rapidly degraded under aerobic conditions. The relationship between anaerobiasis and clogging is shown in Fig. 15.14. Most recharge processes periodically rest infiltration ponds to allow degradation of accumulated polysaccharides and to reaerate the sand column.

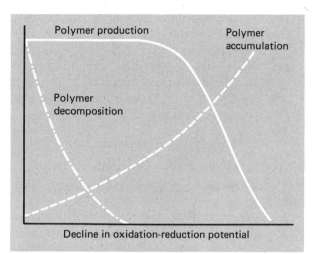

FIGURE 15.14 Clogging of infiltration ponds used for ground-water recharge. When the sand-water interface becomes anaerobic, microbial degradation of bacterial polymers ceases but polymer production continues. The result is a net accumulation of polymers, in the sand interstices, causing clogging.

COMPOST

Municipal waste, raw sewage, or sludge can be treated to yield a stable form of organic matter. The process of composting is a controlled biodegradation of readily available organic compounds. Even cellulosic wastes are degraded.

In developed countries, compost provides a means of recycling municipal solid waste. The Netherlands composts more than 30% of its municipal waste. Underdeveloped countries compost human feces, or *night soil*. The end product is rich in nitrogen and phosphorus. The high temperature of the process destroys most pathogens.

Substrates   The development of a stable compost requires that the organic substrate contain adequate nitrogen.   A carbon:nitrogen ratio of 50 is considered reasonable.   Rubbish is very rich in carbonaceous wastes so that the C:N ratio is too high.   Rubbish mixed with garbage or sewage sludge has a C:N ratio of about 25 and makes excellent compost.   The C:N ratio of rubbish can also be lowered by adding cow manure or wastes from fish or canneries.

The Process   The organic substrate, if it is garbage or municipal waste, must be ground fine to allow aeration and microbial growth. Nonbiodegradable materials, such as cans and other metals, are mechanically removed.

The organic matter is degraded aerobically at a moisture level of about 40%.   Microbial action is inhibited by low moisture content or inadequate aeration.   Normally composting is carried out by thermophilic microorganisms at a temperature above 60°C. This has the advantage of

1. accelerating the process,
2. killing pathogenic microorganisms,
3. destroying ungerminated seeds of weeds.

The microbial population of compost is highly specific.   It consists of an enrichment of aerobic thermophilic bacteria, actinomycetes, and fungi adapted to the decomposition of the substrates commonly used in the compost.   Normally, starter cultures are inoculated from old to new compost fermentations.

Composting Plants   The most primitive method of composting consists of piles of leaves mixed with night soil or agricultural wastes to give a correct C:N ratio. The mixture is turned regularly for aeration.   After some weeks or months, depending on moisture and climate, a stable compost is formed.

Municipal refuse is usually composted in closed digesters (Fig. 15.15).   Commercial digesters are currently in operation in a number of countries.   Most plants operate on the principle of controlled aeration of a ground mixture of wastes.   The temperature, pH, and C:N ratio are all controlled.   Aeration and mixing are achieved by continually moving the material from one level to the other.   Good compost can be obtained in 10 to 30 days. The compost is pasteurized to kill pathogens before sale to the farmer.

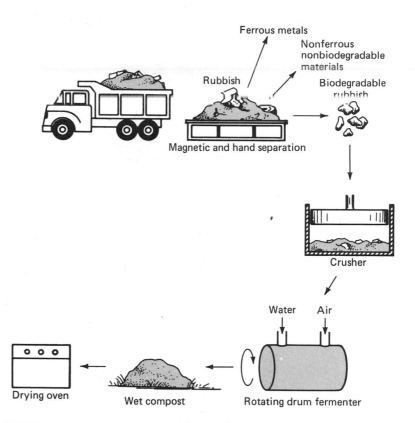

FIGURE 15.15 A schematic diagram of a typical composting plant. Large non-biodegradable materials are removed from the refuse by magnetic and hand separation. The remaining material is crushed and transferred to a rotary drum for fermentation. Water and air are mixed with the refuse. The product of the fermentation process is biologically stable compost. This is transferred to ovens or to windrows to dry.

Agricultural Use   Compost is a stable mixture of organic matter containing nitrogen and phosphorus in a bound form (Fig. 15.16). The addition of compost to soil yields the following results:

1. It improves soil texture so that water-holding capacity and aeration are increased.

2. Nitrogen and phosphorus in the compost are conserved in the organic form. They are released by microbial mineralization only when they are required for crop growth.

3. Added fertilizer is bound more strongly and leaching losses are decreased.

4. Municipal compost provides trace elements including copper, zinc, and molybdenum.

279

FIGURE 15.16 Stable compost formed from municipal refuse. (Courtesy J. Goldstein.)

Commercial inorganic fertilizers are expensive in underdeveloped countries. Composting is used as a means of recycling nitrogen and phosphorus in human and agricultural wastes by using them as soil fertilizers. In more developed countries, composting has the dual function of disposing of municipal wastes and providing a soil conditioner. The addition of compost replaces essential organic matter and prevents erosion in soils depleted of organic matter by intensive agriculture. Compost is also used to reclaim areas where erosion has already occurred. It is particularly useful in the reclamation of land used for mining and in the stabilization of steep slopes.

SUMMARY

1. Biological sewage treatment is used to eradicate pathogens and to oxidize available organic matter. Secondary treatment is the core of the biological process.

2. Oxidation ponds and septic tanks offer an alternative to conventional waste treatment. These methods of waste disposal are common in rural areas.

3. Drinking water is treated to kill pathogens and to remove organic or inorganic colloids and coloration.

4. Land disposal of sewage is practiced as an alternative to waste treatment. The waste water is infiltrated into soil to recharge groundwaters.

5. Solid wastes can be reclaimed by composting. Compost is used as a soil additive to improve agricultural productivity and to maintain soil texture.

FURTHER READING

American Public Works Association, "Municipal Refuse Disposal," Chicago, Ill, 1966.

G. M. Fair and J.C. Geyer, *Water Supply and Wastewater Disposal.* John Wiley & Sons, Inc., New York, N.Y., 1954.

G. M. Fair, J. C. Geyer, and D. A. Okun, *Water Purification and Wastewater Treatment and Disposal.* John Wiley & Sons, Inc., New York, N.Y., 1968.

J. W. M. La Riviere, "A Critical View of Waste Treatment," in *Water Pollution Microbiology*, ed. R. Mitchell. John Wiley & Sons, Inc., New York, N.Y., 1972.

W. J. Weber, *Physico-Chemical Processes for Water Quality Control.* Wiley-Interscience, New York, N.Y., 1972.

J. E. Zajic, *Water Pollution, Disposal and Re-Use*, Vol. 1. Marcel Dekker, Inc., New York, N.Y., 1971.

ADVANCED

WASTE TREATMENT

16

The conventional treatment of wastes removes more than 90% of the available organic matter. However, refractory organic materials are only partially removed (Table 16.1). The short detention time in a secondary treatment plant is insufficient to degrade many complex organic compounds. In addition, the plant only removes 50% of the nitrogen and 30% of the phosphorus in municipal waste. The remainder is mineralized by the microflora in the treatment plant. It is released to the receiving waters as inorganic nitrogen and phosphorus compounds which stimulate algal productivity.

TABLE 16.1   Removal of pollutants by primary and secondary waste treatment.[a]

Parameter	Primary Treatment (%)	Secondary Treatment (%)
Biochemical oxygen demand	35	90
Chemical oxygen demand	30	80
Refractory organics	20	60
Suspended solids	60	90
Total nitrogen	20	50
Total phosphorus	10	30
Dissolved minerals	—	5

[a]From "Cleaning Our Environment—The Chemical Basis for Action," a report by the Subcommittee on Environmental Improvement, Committee on Chemistry and Public Affairs, American Chemical Society, 1969, p. 38. Reprinted by permission of the copyright owner.

In this chapter we consider advanced treatment processes. These are used to take nutrients and recalcitrant organics from waste water.

Many sources of drinking water are high in salinity. Municipalities may be forced to use brackish or reclaimed waters for their drinking water supplies. Advanced water treatment can be used for desalination and removal of hardness or heavy metals from drinking water.

PHOSPHORUS REMOVAL

Phosphate removal from secondary effluent is most easily achieved by precipitation with lime [$Ca(OH)_2$]. Under the alkaline pH conditions of wastewaters, calcium carbonate and hydroxyapatite are precipitated according to reactions 1 and 2:

(1) $Ca(OH)_2 + Ca(HCO_3)_2 \rightarrow 2\,CaCO_3\downarrow + 2\,H_2$

(2) $5\,Ca^{2+} + 4\,OH^- + 3\,HPO_4^- \rightarrow Ca_5\,OH(PO_4)_3 + 3\,H_2O$

The quantity of lime required for phosphate removal is directly related to the alkalinity of the wastewater. Orthophosphate is only converted to an insoluble form at pH values above 9.5. If the lime requirement to raise the pH of the wastewater above 9.5 is too high, then the use of lime for phosphate removal becomes

FIGURE 16.1 Use of lime to remove phosphate from sewage. The coagulation of phosphate is dependent of pH. Significant removal only occurs when the lime increases the pH above 9.5. (From *Advanced Wastewater Treatment* by Culp and Culp. © 1971. Reprinted by permission of Van Nostrand Reinhold Co.)

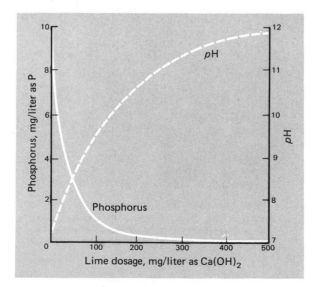

uneconomical. Some typical figures for removal of phosphate from sewage using lime can be seen in Fig. 10.1.

When the alkalinity of wastewater is low, alum [$Al_2(SO_4)_3$] is used to precipitate phosphate. The reaction is

$$Al_2(SO_4)_3 + 2\ PO_4^{2-} \rightarrow 2\ AlPO_4 + 3\ SO_4^{2-}$$

The reaction is strongly pH dependent. It is most rapid at pH values between 6.0 and 7.5. The relationship among alum concentration, pH, and phosphorus removal rate is illustrated in Fig. 16.2.

The precipitated phosphate is sedimented and removed in *tube* settlers, shown in Fig. 16.3. These are shallow tubes that permit rapid settling and removal of the phosphate precipitate.

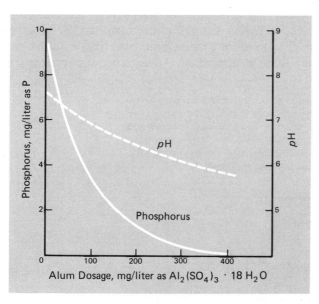

FIGURE 16.2 Phosphate precipitation by alum. This process is commonly used in acidic wastewaters. (From *Advanced Wastewater Treatment* by Culp and Culp. ©1971. Reprinted by permission of Van Nostrand Reinhold Co.)

FIGURE 16.3 Tube settlers are used to separate sediments following coagulation. The shallow tube allows the sludge to settle rapidly with very short detention times, often less than 5 minutes.

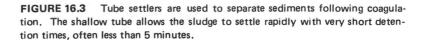
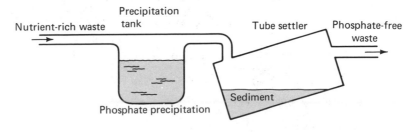

**NITROGEN
STRIPPING**

Nitrogen occurs in wastewaters as ammonia, nitrite, and nitrate. Advanced treatment requires the removal of both nitrogen and phosphorus from wastewaters. Two methods are available for nitrogen removal:

1. ammonia stripping,
2. microbial denitrification.

Ammonia Stripping   The nitrogen in raw waste is almost entirely in the form of ammonia or reduced organic compounds. Secondary treatment normally oxidizes the ammonia to nitrate. By maintaining a high level of organic matter, however, the nitrogen in the secondary effluent can be maintained in the form of ammonia.

Ammonia stripping of the treated secondary effluent is achieved by increasing the pH and temperature. Ammonia is highly soluble at pH 7. When the pH is increased to 10, most of the ammonium ion is converted to ammonia gas. The volatility of the gas is increased with temperature. The ammonia is removed

FIGURE 16.4   An ammonia stripping tower used in the South Tahoe, Calif., advanced treatment facility. The pH of the wastewater is increased to pH 10 and the water is allowed to flow down through an updraft of heated air. The ammonia gas is stripped from the water and carried off in the airflow. (Courtesy Cornell, Howland, Hayes, and Merryfield.)

by passing the alkaline wastewater through highly aerated heated towers (Fig. 16.4). More than 98% of the ammonia can be removed from wastewater using this method.

Denitrification   Nitrogen can be removed from oxidized secondary effluent by conversion of nitrates to volatile nitrogen gas. The denitrification process is heterotrophic and requires a readily available carbon source together with the absence of oxygen. Methanol is an inexpensive substrate that is rapidly metabolized by denitrifying bacteria. The volatilization of nitrogen occurs according to the following reactions:

$$3\ NO_3^- + CH_3OH \rightarrow 3\ NO_2^- + CO_2 + H_2O + 2\ OH^-$$

$$2\ NO_2^- + CH_3OH \rightarrow N_2 + CO_2 + H_2O + 2\ OH^-$$

The process is dependent on the optimal environmental conditions required by denitrifying bacteria. The temperature should be 15°C or higher; the pH should be neutral; 90% removal of nitrogen is possible by denitrification. The process requires fine control, however, which is not easily maintained in treatment facilities.

RECALCITRANT ORGANIC MATTER

The detention time in secondary treatment plants is insufficient to remove many organic compounds that are slowly degraded. These materials are released and accumulate in the receiving waters. The removal of recalcitrant organics is particularly important where industrial effluents are being fed into the treatment plant.

Organic materials are also of importance in drinking waters. They may originate in industrial effluents or may be natural in origin. The presence of phenolic compounds in drinking water supplies imparts an unpleasant taste and odor to the water.

Activated Carbon   Passage through an activated carbon column is the most common means of removing recalcitrant organics from either wastewater or drinking water. The materials typically removed by activated carbon are illustrated in Fig. 16.5. Even pesticides such as DDT can be removed in this way.

Activated carbon has a very large surface:volume ratio and can adsorb large quantities of organic material. It operates optimally

Materials Removed by Activated Carbon:

Pesticides

Industrial wastes

Lignins

Algal toxins

Taste- and odor-producing organics

FIGURE 16.5 Recalcitrant organic materials absorbed from water by activated carbon.

FIGURE 16.6 An activated carbon column used in the South Tahoe, Calif., advanced treatment facility. A contact time of 15-35 minutes removes about 95% of the recalcitrant organic compounds. (Courtesy Cornell, Howland, Hayes, and Merryfield.)

at pH 7. Adsorption is weak below pH 5 and above pH 9. An activated carbon column used in tertiary treatment is shown in the photograph in Fig. 16.6

Ultrafiltration Membranes can be fabricated to act as molecular sieves. *Ultrafiltration* is a process in which large molecules are filtered out of water by passing the water through a membrane under pressure (Fig. 16.7). Membranes are made of polymeric materials with pore sizes tailored to screen out specific organic compounds.

Ultrafiltration is still in its infancy as a process for use in water and wastewater treatment. It should prove useful, however, as a means of removing specific recalcitrant chemicals, particularly from industrial effluents.

FIGURE 16.7 An ultrafiltration cell. Gas pressure forces the water through the membrane, leaving the large organic molecules behind. (Courtesy Amicon Corp.)

DESALINATION

Effluents from secondary treatment plants frequently contain 300 mg/liter more dissolved inorganic salts than the local water supply. In urban districts we are continually recycling our natural waters. We use the same water source repeatedly for waste disposal and drinking water. The salt concentration inevitably rises. In many areas the drinking water source is naturally saline. The salinity can be removed from saline drinking water supplies by two methods:

1. electrodialysis,
2. reverse osmosis.

Distillation is not a reasonable economic possibility unless very large quantities are treated.

Electrodialysis This process removes salts from water by placing it in contact with ion-permeable membranes. The basic electrodialysis apparatus is shown in Fig. 16.8. The feedwater passes

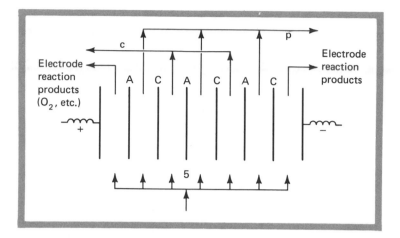

FIGURE 16.8 A diagram showing the electrodialysis desalination process. Anion-exchange (A) and cation-exchange (C) membranes are placed in series. A current is passed across the membranes as the saline water (S) is fed between them. The ions move across the membranes and are removed in the concentrate (c). The product water (p) is free of salts. (After L. H. Shaffer and M. S. Mintz, *Principles of Desalination*, by K. S. Spiegler. Academic Press, Inc., N. Y., 1966.)

between anion- and cation-permeable membranes. When a current is passed across the membranes, the ions in solution move across the appropriate membranes. The product water is desalted leaving two residues, one of cation-rich and the other of anion-rich water. Stacks of membranes are used in the desalination process.

A commercial desalination stack is shown in Fig. 16.9. The

FIGURE 16.9 An electrodialysis stack. (Courtesy Ionics, Inc.)

accumulation of microorganisms and their products on the membranes is a serious problem in electrodialysis plants. The microbial film increases the resistance across the membrane, requiring increased current to maintain ion transport. Ultimately the required current reaches an uneconomical level and the stack of membranes must be dismantled and cleaned.

Electrodialysis has uses other than desalination of water. The process removes metals from effluents. Electroplating wastes can be efficiently removed and recycled by using this process.

Prefiltration before electrodialysis is used to prevent membrane clogging. Growth of microorganisms on the membranes can be prevented by chlorination or by bubbling a fine stream of air along the membrane-water interface.

Reverse Osmosis This process is similar to ultrafiltration. The membrane used is cellulose acetate, however, and the molecules filtered are inorganic salts. The membrane is semipermeable, allowing flow of water but impeding the flow of salts. The flow rate is increased by exerting pressure on the water.

The cellulose acetate membranes used in reverse osmosis are susceptible to biodegradation. The life of a reverse osmosis system is determined by the fragility of the membranes.

Both ultrafiltration and reverse osmosis filtration systems suffer from the accumulation of microorganisms and their products on the membrane surfaces. No method is available to prevent these accumulations. The systems must be periodically dismantled and cleaned.

ION EXCHANGE

Resins can be used to remove ions from water. The principle of ion exchange utilizes the ability of clays or synthetic resins to exchange strongly adsorbed ions for ions in solution. Cation-exchange resins contain acidic groups, such as $-COOH$, carboxyl, and $-SO_3H$, sulfonic. These groups exchange with cations in the solution. Anion-exchange resins contain basic amino groups, such as $-NH_2$, a primary amine group, or $-RNH$, a secondary amine. These amino groups exchange with anions in solution.

The ion-exchange resins are placed in columns and water passes through them.

Principle A typical reaction of a cation-exchange resin would be

$$2\ R-SO_3H\ +\ Ca^{2+} \rightarrow (RSO_3)_2Ca\ +\ 2\ H^+$$

This reaction removes calcium from solution. When the resin becomes saturated with calcium, it is regenerated by passing a strong acid through the column to displace the calcium.

The analogous reaction of an anion-exchange resin would be

$$2\ RNH_3OH + SO_4^- \rightarrow (RNH_3)_2SO_4 + 2\ OH^-$$

Sulfate ions are removed from solution. Saturated anion-exchange resins are regenerated with a strong base.

Uses Ion exchange is primarily used for water softening. Both calcium and magnesium are easily removed by this method. In areas where the water is very hard, it is common to install ion-exchange columns in homes.

It has been suggested that nutrients might be removed by ion exchange in tertiary treatment plants. Anion-exchange resins are capable of efficient phosphate removal. In some cases, more than 95% of the phosphate is removed from the wastes. Similar efficiencies have been obtained in the removal of nitrates, sulfates, detergents, and recalcitrant organic compounds.

Unfortunately it is difficult to predict the composition of secondary effluents from day to day. Ion-exchange resins are

FIGURE 16.10 Tertiary treatment at the South Tahoe, Calif., waste treatment facility. *Foreground*: A secondary clarifying tank for the activated sludge process. *Middle*: A clarifying tank to remove precipitated phosphates. *Background*: Ammonia stripping tower. (Courtesy Cornell, Howland, Hayes, and Merryfield.)

quite specific and require a uniform effluent for efficient operation. The most satisfying results have been obtained when only one material, e.g., phosphate, is removed.

The application of ion-exchange techniques for the removal of a single ion has been exploited in the recycling of electroplating wastes. Copper, zinc, and nickel are removed by cation-exchange resins and chromate is removed by anion-exchange resins.

A TERTIARY TREATMENT PLANT

A complete waste treatment facility handling 7.5 million gallons of sewage per day is being operated at South Tahoe, Calif. (Fig. 16.10). The plant uses advanced treatment to remove phos-

FIGURE 16.11 The South Tahoe, Calif., advanced waste treatment plant. The plant is equipped to handle 7.5 million gallons of waste per day. It receives secondary effluent from an activated sludge treatment facility. Phosphorus, nitrogen, and recalcitrant organic compounds are removed. (From A. F. Slechta and G. L. Culp, *J. Water Pollut. Contr. Fed.,* 39:787 (1967).)

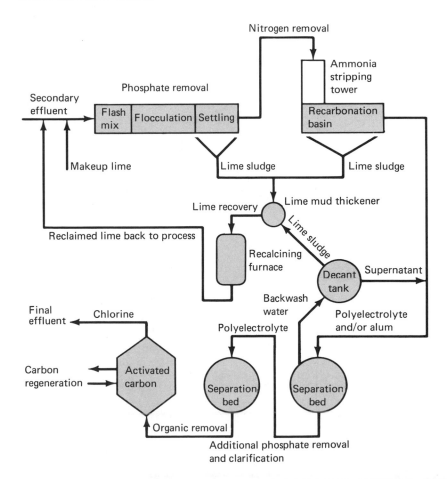

phate, nitrogen, and recalcitrant organic compounds following conventional sewage treatment. The flow diagram of the advanced treatment portion of the plant is illustrated in Fig. 16.11.

Phosphate is removed by the addition of lime. Flocculation of the precipitated phosphate is aided by mixing and polyelectrolyte addition.

Nitrogen is maintained as ammonia in the secondary treatment phase. The ammonia is stripped by passing the high pH effluent from the phosphate treatment through an ammonia stripping tower. At the alkaline pH the ammonia is in the gaseous form and escapes easily with the high temperature and forced aeration in the stripping tower.

The effluent from the nitrogen removal plant is adjusted to pH 7 and passed through granular activated carbon filters to remove recalcitrant organic materials. Contact time is 15 to 45 minutes. The effluent is colorless and odorless.

The efficiency of operation of the South Tahoe plant is summarized in Table 16.2. Tertiary treatment yields an effluent that is not only low in B.O.D., C.O.D., and coliforms but is also deficient in nutrients and recalcitrant organic compounds. It must be noted that the South Tahoe facility does not remove all inorganic salts. In fact, the sulfate level is increased. Electrodialysis or reverse osmosis treatment could be used to remove these ions.

TABLE 16.2   Quality of purified reclaimed water after tertiary treatment at South Tahoe, Calif. The reclaimed water is extremely pure.[a]

Measurement	Wastewater (mg/liter)	Reclaimed Water (mg/liter)
B.O.D.	200-400	< 1
C.O.D.	400-600	3-25
Total organic carbon	—	< 1-7.5
Phosphate	25-30	0.2-1.0
Organic nitrogen	10-15	0.3-2.0
Ammonia	25-35	0.3-1.5
Nitrate and nitrite	0	0

[a]From A. F. Slechta and G. L. Culp, *J. Water Pollut. Contr. Fed.*, **39**:787 (1967).

This plant provides a blueprint for waste treatment in the future. It would not be surprising if biological secondary treatment is ultimately replaced by physiochemical processes.

SUMMARY

1. Advanced waste treatment may be used to remove recalcitrant organic compounds, heavy metals, and nutrients. Saline and hard drinking water sources can be purified using advanced methods.

2. Phosphorus can be removed from eutrophic waters by precipitation. Nitrogen removal is achieved by ammonia stripping or by biological processes.

3. Recalcitrant organic compounds are most commonly removed by adsorption onto activated carbon. Ultrafiltration serves the same purpose.

4. Salts can be removed from drinking water by either electrodialysis or reverse osmosis. Calcium, magnesium and heavy metals can be removed by ion exchange.

5. A complete advanced waste treatment facility has been built at South Tahoe, Calif. The plant removes B.O.D., coliforms, nitrogen, phosphorous, and recalcitrant organics.

FURTHER READING

American Chemical Society, *Cleaning our Environment, the Chemical Basis for Action*. Washington, D.C., 1969. See section on advanced waste treatment.

R. L. Culp and G. L. Culp, *Advanced Wastewater Treatment*, D. Van Nostrand Co., Inc., Princeton, N.J., 1971.

K. S. Spiegler, *Principles of Desalination*. Academic Press, New York, N.Y., 1966.

W. J. Weber, *Physico-chemical Processes for Water Quality Control*. John Wiley & Sons, Inc., New York, N.Y., 1972.

J. E. Zajic, *Water Pollution, Disposal and Re-Use*, Vol. 2. Marcel Dekker, Inc., New York, N.Y., 1971.

MICROORGANISMS

AS

FOOD

The possibility of recycling human, agricultural, and even industrial wastes as a source of food is being investigated in many parts of the world. Human feces has been used as an agricultural fertilizer for centuries in Asia. The rapid rise in population during the past 50 years, particularly in underdeveloped countries, has outstripped our capacity to produce food. The trends in per capita food production throughout the less developed part of the world have been toward a decline in per capita food output (Fig. 17.1). At the same time the developed part of the world is improving its ability to feed itself. This difference reflects the inability of the

FIGURE 17.1 The trend toward a decline in per capita food output in the underdeveloped nations contrasts with the increase in per capita food output in the developed nations. (Reprinted from *Single-Cell Protein* by R. I. Mateles and S. R. Tannenbaum by permission of The M.I.T. Press, Cambridge, Mass., 1968.)

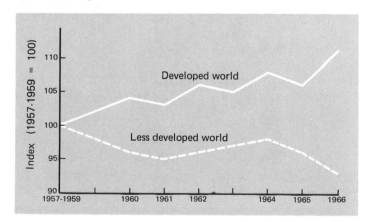

developing nations to increase their food output as rapidly as their population growth. Even in the developed nations the population continues to grow. Ultimately crop yields will reach a maximum and alternative sources of food will be required.

SINGLE-CELL PROTEIN

It is possible to use both inorganic and organic wastes as substrates for specific microorganisms. These microorganisms can be used either as a direct source of protein in food, *single-cell protein*, or as food additives to supply essential growth factors. Alternatively, they can be used to feed domestic animals, providing them with protein and/or growth factors. Yeast makes excellent animal fodder, particularly since it contains 53 to 56% protein. Table 17.1 illustrates the fact that the concentration of essential amino acids for humans produced by brewers' yeast compares well with milk casein and egg protein.

The protein content of other edible microorganisms is equally high. *Lactobacillus fermentans*, which is commonly found in milk, contains 87% protein (Table 17.2). Fungi are very high in protein

TABLE 17.1 A comparison of the essential amino acid content of brewers' yeast, milk casein, and egg protein.

Yeast compares very favorably in all essential amino acids except those containing sulfur.

Amino Acid	Milligrams Amino Acid per Gram Nitrogen		
	Brewers' Yeast	Milk Casein	Egg
Tryptophan	96	84	103
Threonine	318	269	311
Isolcucine	324	412	415
Leucine	436	632	550
Lysine	446	504	400
Total sulfur amino acids	187	218	342
Phenylalanine	257	339	361
Valine	368	465	464
Arginine	304	256	410
Histidine	169	190	150

TABLE 17.2   Protein content of some edible microorganisms.

The bacteria contain the most protein and the algae contain the least.[a]

Microorganisms	Protein Percent of Dry Weight
Bacteria	
Lactobacillus fermentans	87
Escherichia coli	82
Lactobacillus casei	47
Fungi	
Penicillium notatum	38
Yeasts	
Saccharomyces cerevisiae	53–56
Algae	
Chlorella pyrenoidosa	44
Chlorella vulgaris	24–44

[a]H. J. Bunker in *Biochemistry of Industrial Microorganisms,* by L. Rainbow and A. H. Rose. Academic Press London and New York, 1963. Reproduced from R. F. Anderson and R. W. Jackson, *Applied Microbiology,* 6: 369 (1958).

content. *Penicillium notatum* is 38% protein. The green alga *Chlorella pyrenoidosa* contains 44%. The importance of species can be seen by the observation that *Chlorella vulgaris* may contain only 24% protein.

The vitamin content of edible microorganisms is very high indeed. Table 17.3 shows the vitamin B content of brewers' yeast and *Chlorella.* The B vitamins, thiamin, riboflavin, nicotinic acid, panthothenic acid, pyridoxine, biotin, and folic acid, are produced in high concentration by yeast. *Chlorella* lacks only biotin and pyridoxine.

The rapidity of growth of microorganisms make them ideal food sources. One can obtain 1 lb of protein from a 1000-lb bullock in 1 day's growth. In the same period of time, 1000 lb of yeast would produce *50 tons* of protein. This is a function of the much more rapid growth rate of microorganisms.

The short generation times of edible microorganisms can be seen in Table 17.4. *E. coli* grows extremely rapidly. A population of 10 million cells per liter of medium can be obtained in 6 hours.

TABLE 17.3   Vitamin B content of brewers' yeast and *Chlorella*. Yeast produces all the B vitamins. *Chlorella* is incapable of synthesizing biotin or pyridoxine.

	Brewers' Yeast	Chlorella
Thiamine	+	+
Riboflavin	+	+
Nicotinic acid	+	+
Pantothenic acid	+	+
Biotin	+	−
Pyridoxine	+	−
Vitamin B$_{12}$	+	+

Pseudomonas grows half as fast as *E. coli* and yeast grows only at a quarter of the speed. Algae are relatively slow growers compared to bacteria or yeasts. Two to four generations in a day are common.

The criteria for a microorganism to be used as a source of food are shown in Fig. 17.2. The most important criteria are that the organism be nontoxic and grow rapidly on a simple nonspecific medium. It should have a high nutritional or vitamin content and be edible by either humans or domestic animals. The organism should also utilize the energy source efficiently without producing an undesirable effluent. The cells should be easy to separate from

TABLE 17.4   Growth rates of some edible microorganisms.

The rate is expressed for optimal growth conditions. A value of 0.3 is equivalent to a generation time of 1 day.[a]

	Organism	Growth Rate
Bacteria	Escherichia coli	26.0
	Pseudomonas fluorescens	13.0
	Azotobacter chroococcum	6.0
Yeast	Hansenula anomala	6.0
Algae	Chlorella pyrenoidosa	0.85
	Scenedesmus quadricauda	0.88
	Euglena gracilis	0.60

[a]Reproduced with permission from "Physiology of the Algae," *Annual Review of Microbiology*, 5: 170. Copyright by Annual Reviews, Inc., 1951. All rights reserved.

Technical

1. Rapid growth
2. Simple media
3. Suspended culture
4. Simple separation
5. Freedom from infection—stable fermentation
6. Efficient utilization of energy source
7. Disposable effluent

Physiological

8. Capable of genetic modification
9. Nontoxic
10. Good taste
11. Highly digestible
12. High nutrient content
13. Protein, fat, and carbohydrate content of high quality

FIGURE 17.2 The chief factors that are desirable if a microorganism is to be used as a food source. (Reprinted from *Single-Cell Protein,* ed. R. I. Mateles and S. R. Tannenbaum by permission of The M.I.T. Press, Cambridge, Mass., 1968.)

the medium. This final criterion is quite frequently difficult to meet without sophisticated equipment. The expense of separation frequently makes microbial food uneconomical, particularly in developed countries where other protein or vitamin sources are easily available.

BACTERIA

The rapidity of growth and flexibility of substrates makes bacteria ideal candidates for the production of food. Dried cells of *E. coli* containing 82% protein have been fed to chicks and found to be an excellent food additive. Many edible bacteria synthesize large quantities of sulfur amino acids. This makes them particularly attractive as food additives.

Bacterial single-cell protein is produced from hydrocarbon wastes by the petroleum industry in France, Japan, Taiwan, and India. *Pseudomonas* grows well on petroleum products. It yields cells with a protein content of 69%, of which 34% consists of essential amino acids and 35% is nonessential amino acids.

Nitrogen is added to the fermenter as ammonium and aeration

is vigorous. The fermentation is carried out at pH 7 at 35°C with no addition of growth factors.

Pseudomonas is used in Taiwan as a source of single-cell protein. It has been found to have the following attributes:

1. It possesses a high growth rate.
2. It gives a high yield from the raw materials used.
3. The protein content is high.
4. It contains all essential amino acids.
5. No growth factors are required.
6. Fermentation occurs at the high temperatures found in the tropics.
7. The product is tasteless and odorless and is acceptable as human food.

YEAST

Yeast cells are much larger than bacteria and are much easier to harvest. In the alcohol fermentation the yeast is flocculated and settled out of the medium. Yeast extract from the alcohol industry has been used as a food additive for generations.

Protein content averages 40% on most substrates. In addition, yeast is one of the richest sources of human vitamins. It is capable of synthesizing virtually all our vitamin requirements in high concentration.

The potential of yeast as a human food is summarized in Fig. 17.3. Conversion of 100 lb of carbohydrate yields 65 lb of yeast. The same amount of carbohydrate yields 20 lb of pork, 15

FIGURE 17.3 Yeast has a much greater potential for protein production than conventional food sources. (After K. L. Cartwright, Proc. Inter-American Food Cong., 1958.)

Source	Quantity of Protein Produced from 100 lb of Sugar (lb)
Yeast	65
Swine	20
Cattle as milk	15
Poultry	5
Cattle as beef	4

lb of milk, 5 lb of poultry meat, and only 4 lb of beef. It has been suggested that a percentage of sugar-producing plants be used to produce yeast for use as human food.

Industrial Wastes   More than 125,000 tons of yeast are produced in Europe annually. Most of it is used to feed livestock. The substrates are usually industrial waste materials. Wood hydrolysates and beet molasses are the most common substrates.

In the United States, 69,000 tons of dried yeast was produced in 1963. The main substrate was molasses and most of the yeast was used for baking bread. Spent sulfite liquors from the wood pulp industry have also been used in the United States to produce yeast with some success.

Hydrocarbons   *n*-Paraffins are excellent substrates for yeasts. Growth is rapid in well-aerated cultures and the yield of protein and essential amino acids is excellent. Table 17.5 summarizes the characteristics of yeast produced by fermentation of *n*-paraffins.

The cells contain 66% protein. The full range of amino acids are present in the protein. Fat content is low and the digestibility is very high. Lysine and methionine content of the yeast is very high. These amino acids are essential for humans and are often present in low concentrations in even high protein meat diets.

Figure 17.4 illustrates the process used by British Petroleum Co. Ltd. for the production of single-cell protein from *n*-alkanes. Mineral salts are added together with the hydrocarbon to the fermentation tank. Following fermentation, the cells are separated

TABLE 17.5   **Characteristics of food produced by growing yeast on *n*-paraffin hydrocarbons.[a]**

Characteristic	Quantity
Protein content	66%
Lipid content	0.5%
Amino acid content	19 essential amino acids
Digestibility	90%

[a]Reprinted from *Single-Cell Protein* by R. I. Matales and S. R. Tannenbaum by permission of The M.I.T. Press, Cambridge, Mass., 1968.

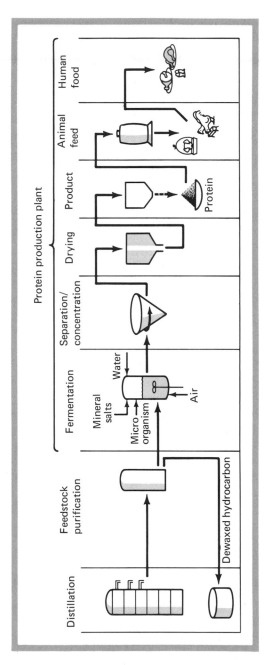

FIGURE 17.4 The process used by British Petroleum to produce single-cell protein from hydrocarbons. *n*-Alkanes are distilled for the use in the fermenter. Minerals are added. Following fermentation, the cells are separated and dried for use as animal feed. (Courtesy British Petroleum Co. Ltd.)

FIGURE 17.5 A plant used by British Petroleum for the production of single-cell protein. (Courtesy British Petroleum Co. Ltd.)

from the liquid and dried for use as animal feed. A British Petroleum plant is shown in Fig. 17.5.

ALGAL NUTRIENTS

I have discussed the problems of eutrophication in Chapter 11. Large quantities of inorganic nutrients are released both from secondary treatment plants and from livestock feedlots. These nutrients would make an ideal fertilizer for the production of algal protein, particularly in warm climates with high light intensities. In this way nutrients could be recycled to produce a food source.

The Process Domestic sewage is a particularly good substrate for algal growth. Algae can be produced in large quantities in shallow artificial ponds fed with sewage (Fig. 17.6). Heterotrophic bacteria degrade the organic matter in the sewage to provide a balanced mineral diet for the algae. The algae utilize the inorganic products of bacterial growth. Agricultural or industrial wastes can

FIGURE 17.6 Use of domestic sewage for algal culture. (Courtesy California Department of Water Resources.)

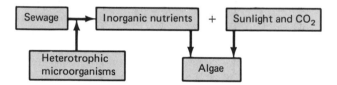

FIGURE 17.7 Domestic sewage is an excellent source of inorganic nutrients. Algal culture has been used successfully in large ponds containing sewage. The algae grow on the inorganic nutrients produced during mineralization of the organic wastes by a heterotrophic microflora. The process is only applicable in warm climates with high levels of sunlight.

be substituted for sewage provided there is sufficient nitrogen and phosphorus present in the waste.

The process is illustrated in the flow diagram in Fig. 17.7. The culture is mixed continually. This serves to maintain aeration and keep the algae from settling out of the euphotic zone. Mixing must be slow, since rapid mixing prevents illumination of the algae.

Centrifugation is the most rapid and simple means of harvesting algae. However, the energy cost is high. A more realistic approach depends on autoflocculation, which occurs in shallow ponds during periods when the pH rises above 9.5. The flocs

FIGURE 17.8 The harvested algae are allowed to dry in the sun. (Courtesy California Department of Water Resources.)

settle out and can be recovered in a manner similar to sludge collection in a secondary treatment plant. The product is easily dried in the sun (Fig. 17.8).

Nutrition The predominant algae that develop in sewage ponds are usually either *Chlorella* or *Scenedesmus*. These algae require degradation of the cellulosic cell wall to be digestible by humans. However, they are acceptable to livestock. They have a protein concentration of about 50%, and an acre of land can produce 20 tons dry weight of algae per year. This yield is 10 to 15 times higher than soybeans and 25 to 50 times higher than corn.

Spirulina This blue-green alga, shown in Fig. 17.9, grows on the surface of ponds and is very rich in protein. More than 60% of the cell is protein. It has been used in Chad as food for centuries. It

FIGURE 17.9 *Spirulina*, a blue-green alga used as a source of protein. (Courtesy G. Clement.)

grows well at pH values between 8.5 and 11 and has a temperature optimum of 30°C. The potential for recycling wastes in warm climates using this alga is very great. It is easily harvested, extremely nutritious, and quite edible.

SUMMARY

1. Human and agricultural animal feces as well as industrial wastes are excellent substrates for the production of single-cell protein.

2. Bacteria grow well on organic wastes. They have a high protein content and many are acceptable either as food or as food additives.

3. Yeasts are easier to harvest and have an extremely high nutritional content. Yeast is produced in Europe for livestock feed.

4. Effluents from sewage treatment plants and from livestock feedlots are rich in algal nutrients. In warm climates with high light intensity these nutrients can be used to produce algal protein.

FURTHER READING

H. J. Bunker, "Microbial Food" in *Biochemistry of Industrial Microorganisms* by C. Rainbow and A. H. Rose. Academic Press, New York, N.Y., 1963.

R. I. Mateles and S. R. Tannenbaum, *Single-Cell Protein.* MIT Press, Cambridge, Mass., 1968.

MICROORGANISMS

AND

AIR POLLUTION

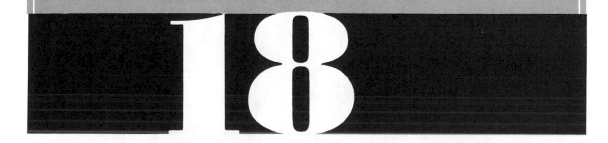

Urban areas are characterized by an accumulation of pollutants in the atmosphere. The high levels of air pollution reached in some cities in the United States can be seen pictorially in Fig. 18.1.

FIGURE 18.1 Two types of air pollution. (a) Smog caused by automobiles in Los Angeles, Calif. (Courtesy U. S. Public Health Service.) (b) An example of industrial emissions of air pollution in New York. (Courtesy U. S. Public Health Service.)

(a)

The ecological effects of air pollution are only beginning to be understood. Both humans and animals are adversely affected. In addition to eye, throat, and bronchial irritation, many pollutants are carcinogenic. Photosynthesis is inhibited by a number of different air pollutants resulting in death of green plants in polluted areas. Microbial processes are involved in both production and neutralization of air pollutants. Microorganisms may be involved in human disease as secondary invaders following pollution induced irritation. Some microorganisms can be used as indicators in the study of cell damage by air pollutants.

CHEMICAL PROCESSES

In the urban environment the composition of the air is altered radically by products of fossil fuel consumption. Table 18.1 shows that automobiles are the primary source of air pollution in the United States. Their emissions contribute 60% of all air pollutants. Industrial and power plant emissions contribute almost equal quantities to make up an additional 30%. Space heating and refuse burning account equally for 10% of the total emissions. The major pollutants in the United States are carbon monoxide, carbon

(b)   **FIGURE 18.1** (continued)

TABLE 18.1 Sources of air pollution in the United States for 1965.
Note that automobiles contribute 60% of all emissions.[a]

Source	Millions of Tons	Percentage of Total
Automobiles	86	60
Industry	23	17
Electric power plants	20	14
Space heating	8	6
Refuse disposal	5	3
Total	142	100%

[a]From *The Sources of Air Pollution and Their Control,* U.S. Public Health Service Report, No. 1548, Washington, D.C., 1966.

dioxide, sulfur and nitrogen oxides, hydrocarbons, and particles. Detailed information on all of these pollutants except carbon dioxide is shown in Table 18.2.

Carbon Monoxide   The burning of fossil fuels by automobiles produces large quantities of carbon monoxide. Local concentrations may be as high as 100 ppm.

TABLE 18.2 Emissions of air pollutants in the United States and their sources for 1965.

Automobiles are responsible for most of the hydrocarbons and nitrogen oxides. Industrial and electric power plants account for much of the sulfur oxides.[a]

Sources	Pollutants (millions of tons)				
	Carbon Monoxide	Sulfur Oxides	Hydro-carbons	Nitrogen Oxides	Particles
Automobiles	66	1	12	6	1
Industry	2	9	4	2	6
Electric power plants	1	12	1	3	3
Space heating	2	3	1	1	1
Refuse disposal	1	1	1	1	1
Totals	72	26	19	13	12

[a]From *The Sources of Air Pollution and Their Control,* U.S. Public Health Service Report, No. 1548, Washington, D.C., 1966.

Carbon Dioxide   Carbon dioxide is produced from combustion of fuels and in heterotrophic respiration. The gas is absorbed in plant photosynthesis. Industrial emissions have increased 15 times within the past 60 years. Plant productivity has not increased at anything like an equivalent rate.

It has been postulated that the accumulation of carbon dioxide in the atmosphere will lead to a "greenhouse effect." This effect would cause the global temperature to rise because of the passage of shortwave irradiation from the sun through the carbon dioxide layer. The carbon dioxide absorbs infrared energy and reemits it back to earth. Thus the amount of radiant energy absorbed by the earth in the presence of high concentrations of atmospheric carbon dioxide would be increased significantly. It has been estimated that for each 10% increase in carbon dioxide concentration the atmospheric temperature would rise 0.5°C.

Despite the fifteenfold increase in carbon dioxide emissions within the past 50 years there has been no significant rise in global temperature. This is probably the result of absorption of excessive atmospheric carbon dioxide into the oceans.

Sulfur Oxides   Combustion of high sulfur fuels, smelting, and oil refinery operations emit both hydrogen sulfurs and sulfur oxides to the air (Table 18.2). These are chemically oxidized to sulfur dioxide and trioxide. These gases are dissolved in water droplets to form sulfuric acid, which either remains in the atmosphere or returns to earth as acid rain.

Hydrocarbons   Automobile exhausts emit 12 million tons of hydrocarbons annually in the United States (Table 18.2). Areas such as Los Angeles, Calif., frequently have concentrations as high as 0.15 ppm, a level at which eye irritation and plant damage occurs.

Industry emits large quantities of hydrocarbons to the atmosphere. Table 18.2 tells us that industrial emissions account for 4 million tons of hydrocarbons annually in the United States. The petroleum industry is the prime industrial source.

Nitrogen Oxides   We can see from Table 18.2 that emissions from automobiles account for almost 50% of all nitrogen oxide pollution. Power plants emit approximately 25%. The remainder is divided among industry, space heating, and refuse disposal.

Atmospheric nitrogen is converted in the combustion process to yield nitric oxide and nitrogen dioxide. Nitric oxide is converted by ozone to nitrogen dioxide. The reaction of nitrogen

NO_2 Nitrogen dioxide	+	light	→	NO Nitric oxide		+	O Atomic oxygen
O	+	O_2 Molecular oxygen	→	O_3 Ozone			
O_3	+	NO	→	NO_2		+	O_2
O	+	Hc Hydrocarbon	→	HcO Radical			
HcO	+	O_2	→	HcO_3 Radical			
HcO_3	+	Hc	→	aldehydes, ketones, etc.			
HcO_3	+	NO	→	HcO_2 Radical		+	NO_2
HcO_3	+	O_2	→	O_3		+	HcO_2
HcO_x Radical	+	NO_2	→	peroxyacyl nitrates			

FIGURE 18.2   A simplified scheme showing the chemical reactions leading to the formation of smog. The catalyst is sunlight and the reactions are photochemical. (From "Cleaning our Environment—The Chemical Basis for Action," a report by The Subcommittee on Environmental Improvement, Committee on Chemistry and Public Affairs, American Chemical Society, 1969, p. 108. Reprinted by permission of the copyright owner.)

dioxide with sunlight is *photochemical* and yields ozone and peroxyl radicals (Fig. 18.2).

Ozone   Ozone is a membrane irritant and causes damage to plants. Both the ozone and the peroxyl radicals react with hydrocarbons to produce the yellowish gas "smog." The series of chemical reactions leading to smog formation are summarized in Fig. 18.2. Unsaturated hydrocarbons are more reactive than saturated ones. Smog is composed of peroxyacyl nitrates (PAN), aldehydes, and ketones, as well as large quantities of ozone.

Particles   Colloidal particles emitted from industrial processes and power plants reached 9 million tons in the United States in 1965 (Table 18.2). In industrial areas, particle accumulation in the air reduces the amount of sunlight and possibly is responsible for lowering global temperatures. Accumulation of particles in the lungs may contribute to respiratory diseases. The aesthetic effect of accumulation of particles on buildings and clothing in the urban environment is unmeasurable but very significant.

MICROBIAL PROCESSES

Most air pollutants originate directly from human activities. However, significant amounts of some pollutants originate either as a result of natural microbial processes or during degradation of wastes produced by human activities. Pollutants may also be scavanged from the air by microbial activities and removed to inactive sinks in the oceans or soil.

Carbon Monoxide It has been calculated that the world's oceans produce between 3 and 10 million tons of carbon monoxide per year. The source has not been identified but it is either marine bacteria or algae. This figure is small compared to the 66 million tons produced in the United States alone by automobiles (Table 18.2).

The absence of high concentrations of carbon monoxide in the global atmosphere indicates that there is an effective scavenging process. Many microorganisms are capable of utilizing carbon monoxide as a carbon source and act as scavengers. The soil microflora may utilize sufficient carbon monoxide to balance both the pollution and the natural load.

Carbon Dioxide There have been immense increases in the CO_2 emissions from human activities during the past 50 years. Yet the concentration of CO_2 in the atmosphere has not risen appreciably. The apparent reason is the ability of the oceans to act as CO_2 sinks. Atmospheric CO_2 dissolves in the oceans where it is utilized by the marine algae. A high percentage of the algal carbon ultimately settles to the ocean floor and is permanently immobilized as calcium carbonate. The oceans have the capacity to dissolve very high concentrations of atmospheric CO_2 and are capable of providing a strong buffering action against CO_2 pollution.

Sulfur Gases Microbiological degradation of organic matter under anaerobic conditions yields high concentrations of hydrogen sulfide. It has been estimated that ocean sediments produce 30 million tons of H_2S annually and that the annual production of H_2S from decaying vegetation is 68 million tons. By comparison, pollution accounts for only 3 million tons of global atmospheric H_2S.

In contrast, sulfur oxide emissions are almost entirely manmade. There are no significant quantities of sulfur oxides released to the atmosphere by microbial activities.

Sulfur dioxide is absorbed from the atmosphere by plants and microorganisms. The sulfur is fixed in the organism's tissue in sulfur amino acids and other sulfur-containing molecules. The biota acts as a sink for an unknown portion of pollution-produced SO_2.

Both H_2S and SO_2 are ultimately converted to sulfates in the

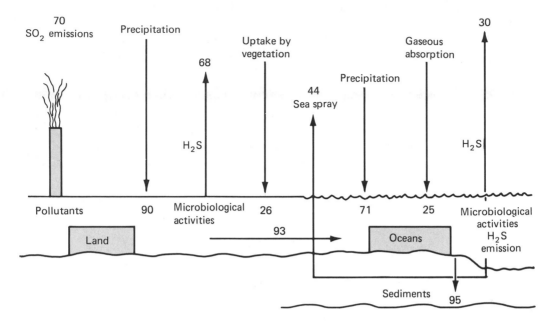

FIGURE 18.3 The circulation of sulfur among the atmosphere, land, and oceans. The numbers are in millions of tons circulated annually. (After E. Robinson and R. C. Robbins, *Air Pollution Control, II,* ed. W. Strauss. John Wiley & Sons, Inc., N. Y., 1972.)

atmosphere. Most of these sulfates are reabsorbed by the oceans because of the high solubility of sulfates in seawater.

The complex circulation of sulfur within the air, land, and oceans is summarized in Fig. 18.3.

Hydrocarbons   Anaerobic processes in swamps produce large quantities of methane. Global swamp methane emissions are probably close to 2000 million tons annually. The calculated residence time in the atmosphere is between 1 and 4 years.

The methane sink is almost certainly microbial. Methane-utilizing bacteria are common in the soil and would be expected to oxidize atmospheric methane. The residence time of methane in the atmosphere probably reflects the numbers and activity of these bacteria.

Nitrogen Oxides   Nitrous oxides are produced microbiologically by denitrification of organic compounds (p. 177). It has been calculated that soil microorganisms produce 10^{15} g of N_2O annually. This is balanced by the uptake of a similar quantity of N_2O by plants and microorganisms.

Nitrogen dioxide (NO_2) is usually formed in the atmosphere by the reaction of nitric oxide (NO) with ozone. Denitrification processes frequently yield NO, although no quantitative data are available for global microbial production. Relative to pollution sources, natural emissions probably are insignificant.

Nitrogen dioxide is scavenged from the atmosphere by chemical processes. No significant biological conversions have been detected.

EFFECTS ON HUMANS

Air pollutants are irritants. Oxides of sulfur cause bronchial irritation. Ozone causes lacrymation. Photochemical smog produces both bronchial irritation and lacrymation.

The disastrous London smog in 1952 caused more than 4000 deaths. Most of these people suffered from respiratory or cardio-vascular disease and the additional stress of the smog killed them. Individuals suffering from these diseases are placed in a hazardous situation by high levels of air pollution.

Many air pollutants are carcinogenic. The incidence of lung cancer among nonsmokers in urban industrial areas is far higher than in rural areas. The carcinogenic hydrocarbons 3,4-benzpyrene and 1,2,5,6-dibenzanthracene are common carcinogenic hydro-carbons found in polluted air. Their structures are shown in Fig. 18.4. The 3,4-benzpyrene concentration in urban districts of the state of New York averages 4 nanograms/cu m of air compared to 0.22 nanograms in a rural district. Table 18.3 summarizes the concentration of 3,4-benzpyrene in some major cities of the world. The low concentration of 3,4-benzpyrene in Oslo and Moscow reflects the low density of automobiles.

Particles increase the incidence of lung cancer. Apparently they absorb carcinogenic chemicals and also irritate the lungs.

FIGURE 18.4 Two carcinogenic polyaromatic hydro-carbons found in air.

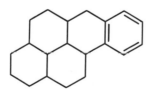

1,2,5,6-dibenzanthracene

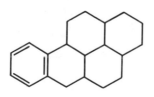

3,4-benzpyrene

TABLE 18.3   Concentrations of 3,4-benzpyrene found in the air in some major cities.

City	3,4-Benzpyrene Concentration (μg/cu m of air)
St. Louis, Mo.	54
London, England	46
Chicago, Ill.	15
Oslo, Norway	0.8
Moscow, U.S.S.R.	0.2

Asbestos has been shown to increase the incidence of lung cancer in smokers but not in nonsmokers.

Exposure to air pollutants predisposes individuals to deep lung microbial infections. Both bacterial and viral infections are increased in the presence of pollutants. Figure 18.5 illustrates the effect of ozone on the infection of mice with an aerosol of *Streptococcus*. The rate of kill of the bacterium in the lung declines. The lag phase before growth is shortened. Exponential growth and the development of septicemia and death occurs more rapidly. Similar results have been found when squirrel monkeys were treated with aerosols of influenza virus together with nitrogen dioxide.

Bronchitis-emphysema is a chronic lung disease that appears to be accentuated by air pollutants. Figure 18.6 explains the pro-

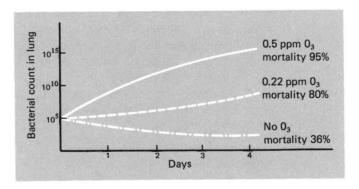

FIGURE 18.5   Exposure of mice to an aerosol of Streptococci following exposure to ozone. Ozone causes an increase in bacterial growth in the lung and a consequent increase in mortality from streptococcal infection. (From D. L. Coffin and E. J. Blommer, Proc. 3rd Internat'l. Symp. Aerobiology, Academic Press, London, 1969.)

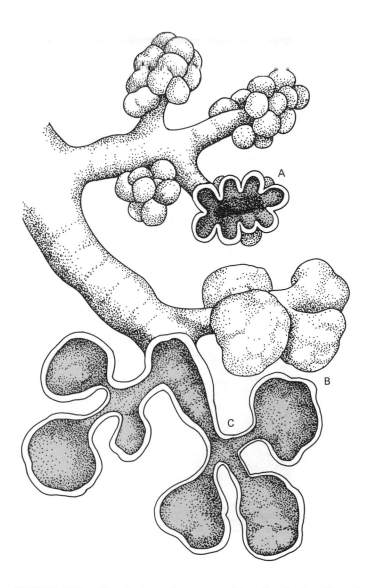

FIGURE 18.6 Chronic lung disease caused by air pollution. Normal lungs have millions of alveoli (A) that transfer oxygen to the bloodstream. When the lung becomes diseased, the alveoli walls deteriote (B) causing a reduction in the amount of membrane available for oxygen transfer. The smallest branches of the bronchial tree also narrow (C) further reducing oxygen exchange. (From W. McDermott, "Air Pollution and Public Health. "Copyright © October 1961 by Scientific American, Inc. All rights reserved.)

gression of lung disease. In the normal lung, air moves through millions of tiny cells or alveoli that transfer oxygen to the blood. When the lung becomes diseased, the alveoli deteriorate so that there are fewer membranes to transfer oxygen. At the same time the branches of the bronchial tree become narrower so that oxygen transfer is restricted even further.

PLANT DAMAGE

The most obvious effect of air pollutants on plants is the accumulation of particulates on the plant surface. This prevents the access of light and oxygen to the plant. The acid rain produced from oxides of sulfur is highly detrimental to plants, reducing growth rates and killing leaf cells. Significant leaf damage occurs at 1.5 ppm of sulfur dioxide.

Ozone injures plant leaves by oxidizing the leaf cells. The leaves turn a characteristic silver color. Unsaturated hydrocarbons such as benzene or heptene cause a similar effect. PAN and other smog oxidation products are equally damaging to plant cells.

FIGURE 18.7 Bioluminescence of bacteria can be used as a tool to determine damage due to air pollution.

MICROBIAL INDICATORS

Toxicity Many microorganisms are sensitive to air pollutants. These organisms can be used as indicators or for the study of cytological damage. *Escherichia coli* is highly sensitive to smog produced by a photoreaction of ozone and hydrocarbons. The mixture is lethal for *E. coli* in parts per billion. Ozone alone is also toxic for *E. coli*. The cells are destroyed by oxidation of the cell surface, resulting in leakage of the contents.

Luminescent bacteria make excellent tools for the determination of cytological damage caused by air pollutants. They glow in the dark and their bioluminescence is easily determined. The luminescent character of these bacteria is illustrated in Fig. 18.7, which is a photograph of the bacteria taken in total darkness.

Smog produced by photochemical interaction of butene and nitric oxide causes a significant decline in bioluminescence. The bioluminescent bacteria are particularly sensitive to PAN. Concentrations of less than 2 μliters/liter prevent luminescence. No eye irritation can be detected at such low levels of smog. Subtle physiological effects such as these presumably also occur in higher organisms at low air pollution levels. The microbiological bioassay should give an insight into the nature and extent of cell damage caused by air pollutants.

Carcinogens Carcinogenic polycyclic hydrocarbons are common air pollutants. They also stimulate the formation of aberrations in bacterial cells. This phenomenon can be used to study the levels of pollutants necessary to cause cell damage and the nature of the

FIGURE 18.8 Effect of carcinogens on the cell formation of *Bacillus megaterium*. Large atypical cells full of granules form when the bacterium is cultured in the presence of 3,4-benzypyrene. (After W. D. Won and J. F. Thomas, *Applied Microbiology*, 10:217 (1962).)

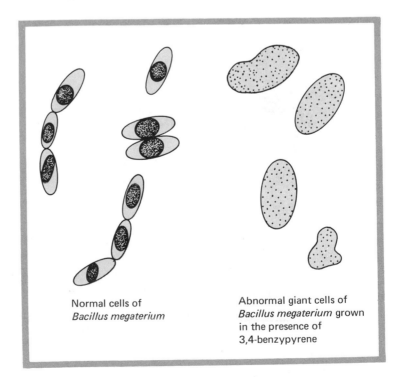

Normal cells of
Bacillus megaterium

Abnormal giant cells of
Bacillus megaterium grown
in the presence of
3,4-benzypyrene

damage. *Bacillus cereus* cells treated with the carcinogenic pollutant 3,4-benzpyrene increase their metabolic activity and abnormal bacteria form. Figure 18.8 shows large granulated cells of *Bacillus megaterium* formed under the influence of benzpyrene.

The toxicity of benzpyrene for microorganisms is accentuated by exposure to ultraviolet light. A protozoan, *Paramecium*, has been used in studies of the effect of light on hydrocarbon toxicity. The process is one of photooxidation of the pollutant since it does not occur in the dark. This *photodynamic response* has been correlated with carcinogenic activity. Epstein and his co-workers tested 50 carcinogenic hydrocarbons and 67 noncarcinogenic hydrocarbons for photodynamic activity using *Paramecium* as the test organism. The results showed that carcinogens are much more photodynamic than noncarcinogens.

Paramecium can be used as a bioassay for carcinogenicity caused by photodynamic responses of hydrocarbons that are air pollutants.

FIGURE 18.9   A normal growth of lichens.

Photosynthesis   Lichens (Fig. 18.9) are excellent tools for the study of the effects of air pollutants on green plants. The symbiotic nature of the lichen increases its sensitivity to stress. Adverse effects can be seen at pollution levels where damage is not apparent in green plants.

Figure 18.10 depicts the development of a lichen desert in the region of Newcastle, an industrial city in England. The number of lichen species observed declines from 55 down to 5 close to the city. The major pollutants in the atmosphere around Newcastle are sulfur oxides. Concentrations close to the city average 200 μg/cu m of air. Industry in the area consumes 200,000 tons of coal

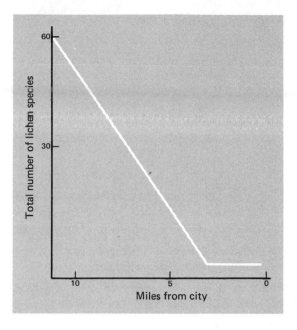

FIGURE 18.10 The development of a lichen desert in the industrial city of Newcastle, England. The number of species declines close to city as the concentration of sulfur dioxide increases. (From O. L. Gilbert, *Ecology and the Industrial Society,* by G. T. Goodman, R. W. Edwards, and J. M. Lambert. Blackwell Scientific Pub. Ltd., Oxford, England, 1965.)

annually. The lichens concentrate sulfur in their tissue. The sulfur content of the lichens in the rural area 21 miles outside the city contained 225 ppm of sulfur (Fig. 18.11). Those sampled 4 miles outside the city contained 2879 ppm of sulfur.

FIGURE 18.11 The sulfur content of the lichen *Parmelia saxatilis* in rural and industrial urban areas close to Newcastle, England. The sulfur content of the air 8 miles from the city was 0.014 ppm. The air contained 0.02 ppm sulfur 4 miles from the city. (From O. L. Gilbert, *Ecology and the Industrial Society,* by G. T. Goodman, R. W. Edwards, and J. M. Lambert. Blackwell Scientific Pub. Ltd., Oxford, England, 1965.)

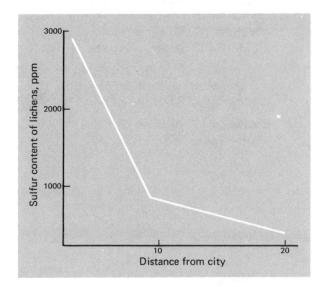

SUMMARY

1. Automobile emissions account for 60% of all air pollution. Power plant and industrial emissions are responsible for a further 30%.

2. The major pollutants are carbon monoxide, carbon dioxide, sulfur oxides, hydrocarbons, and nitrogen oxides.

3. A small quantity of air pollutants is produced by microbiological activities. Microorganisms can scavenge pollutants from the air.

4. Air pollutants may be irritants and/or carcinogens or they may predispose people to deep lung microbial infection.

5. Microorganisms can be used as indicators of air pollution. Potential carcinogenic air pollutants can be screened using bacteria and protozoa.

6. Photosynthetic inhibition of air pollutants can be studied using lichens.

FURTHER READING

D. L. Coffin in *Advances in Environmental Sciences and Technology*, Vol. 2, by J. N. Pitts, Jr. and R. L. Metcalf. John Wiley & Sons, Inc., New York, N.Y., 1971.

W. L. Faith and A. A. Atkisson, Jr., *Air Pollution*, 2nd ed. John Wiley & Sons, Inc., New York, N.Y., 1972.

O. L. Gilbert, "Lichens as Indicators of Air Pollution in the Tyre Valley" in *Ecology and the Industrial Society* by G. T. Goodman, R. W. Edwards, and J. M. Lambert. Blackwell Scientific Pub., Ltd., Oxford, England, 1965.

W. McDermott, "Air Pollution and Public Health," in *Man and the Ecosphere*. Readings from Scientific American, W. H. Freeman and Co., San Francisco, Calif., 1971.

W. Strauss, *Air Pollution Control*, 2 vol. John Wiley & Sons, Inc., New York, N.Y., 1972.

GLOSSARY

Abiotic: Nonliving. As opposed to biotic. Abiotic factors controlling microbial activity include pH and temperature.

Activated carbon: A form of powdered carbon used to adsorb recalcitrant organic compounds from wastewaters as well as tastes and odors from drinking water.

Activated sludge: The mixed microbial population in the activated sludge treatment process for sewage treatment.

Adenosine triphosphate (ATP): A molecule composed of adenosine and three phosphate groups. The phosphates are bound by high energy linkages and are associated with energy transfer in living cells.

Aerobe: An oxygen-requiring organism.

Ammonia stripping: A process for the removal of ammonia from wastewater.

Anabolism: The biochemical processes in which energy is consumed by the cell. At the same time biosynthesis occurs.

Anaerobe: An organism that does not require atmospheric oxygen.

Anoxic: An environment devoid of oxygen.

Antagonism: Destruction or prevention of growth of one organism by another.

Antibiotic: An organic chemical produced by one microorganism that kills or inhibits another microorganism.

Aquifier: The place in the ground where groundwater is naturally stored.

Ascospores: The spores produced by Ascomycetes.

Ascus: A container for ascospores produced by Ascomycetes.

Aseptic: Procedure that maintains sterility.

Autecology: The study of single individuals in relation to ecological processes. In contrast to *synecology*, which is the study of whole communities in relation to their environment.

Autocatalytic: Self-catalyzing. Many chemical reactions are autocatalytic.

Autoclave: A high-pressure steam sterilizer.

Autolysis: Self-destruction of cells by the action of autolytic or intracellular enzymes.

Autotrophic organisms: Those organisms that utilize light energy or the oxidation of inorganic chemicals as their sole energy source.

Axenic: Growth of a microorganism alone in a medium without other organisms being present.

Bacillus: A rod-shaped bacterium with specific physiological characteristics.

Bactericidal: Able to kill bacteria.

Bacteriophage: A virus that infects bacteria.

Bacteriostatic: Capable of preventing bacterial growth but not of killing the bacteria.

Barophile: A microorganism that lives at high pressures, usually in the depths of the ocean.

Basidiospore: A sexual spore produced by the Basidiomycetes.

Basidium: The specialized part of the mycelium of Basidiomycetes on which basidiospores are formed.

Benthos: The shallow-water bottom organisms of the sea.

3,4-Benzpyrene: A strongly carcinogenic aromatic hydrocarbon.

Binary fission: Division of a cell into two daughter cells. The process by which bacteria reproduce.

Biochemical oxygen demand (B.O.D.): The amount of oxygen required to degrade organic matter present in a sample, usually held in the dark at 20°C for 5 days.

Biological control: The direct use of living organisms or their products to control other organisms that have become pests.

Bioluminescence: Light energy produced by luminescent organisms.

Biomass: The mass of an organism present in an ecosystem at any time. Contrast *productivity*, which is the rate of production of that organism.

Biosphere: The portion of the earth that is inhabited by living organisms.

Buffer: A chemical used to prevent changes in pH.

Capsule: A layer of mucoid material surrounding a bacterial cell.

Carcinogenic agent: A material that causes cancer. Compare *mutagenic*, causing mutations, and *teratagenic*, causing birth defects.

Catabolism: The biochemical processes of a cell that generate energy for use by the cell. At the same time substrates are degraded.

Chemoautotroph: An organism that utilizes oxidation of inorganic chemicals for its energy and growth.

Chemostat: An apparatus used to grow bacteria continuously in a specific growth phase.

Chemotaxis: Movement of a cell in response to a chemical attractant or repellent.

Chemotherapy: Control of disease by treatment with chemicals.

Chlamydospore: Fungal spores formed by separation of the hyphae.

Chlorophyll: A green pigment in photosynthetic organisms that acts as an electron donor in the photosynthetic process.

Chloroplast: The portion of eucaryotic and plant cells where photosynthesis occurs.

Cilia: Hairs on certain organisms that are responsible for motility.

Climax community: The mature, stable community developed after many stages of ecological development.

Clone: A population of cells formed from one cell.

Coagulation: An aggregation or flocculation of cells.

Coccus: A spherical bacterium.

Coencytic organisms: Organisms that consist of protoplasm containing many nuclei. Slime molds are coenocytic at one stage in their development.

Coliphage: A virus that infects *E. coli.*

Colony: A visible growth of a protist on a solid nutrient medium.

Commensalism: An association between two organisms that benefits one or both organisms.

Community: A group of populations living closely together as an association.

Composting: A process for decomposing solid wastes to a stable product for use in agriculture.

Conidiophore: A hypha of a fungus that produces conidia.

Conidium: An asexual spore bag produced by Fungi Imperfecti, and containing conidiospores.

Conjugation: The combination of two sexual forms of an organism to yield a new individual.

Consumer: An organism that grows and obtains its energy by utilizing the organic materials of other organisms, living or dead. Contrast *producers*, which use light or inorganic energy.

Contaminate: Introduce external microorganisms to a sample that is either sterile or under controlled growth conditions.

Culture: A defined microbial population growing in a nutrient medium.

Cybernetics: The study of control. Increasingly utilized in ecology.

Cytochrome: Essential electron carriers utilized in respiratory processes.

Denitrification: Enzymatic reduction of nitrates by bacteria to nitrogen gas.

Deoxyribonucleic acid (DNA): A nucleic acid found in most organisms.

Detergents: Cleaning agents. Detergents may be soft, easily biodegraded, or hard, resistant to biodegradation.

Detritus: Dead organic material.

Dialysis: Use of a semipermeable membrane to separate water from dissolved materials.

Disinfectant: A chemical that kills microorganisms.

Diversity: The number of species that lives together in an ecosystem.

Ecology: The branch of science dealing with the relationship between organisms and their environment.

Ecosystem: A community together with its environment.

Electrodialysis: The removal of ions from water by transfer across a membrane using electrical current. Particularly useful in desalination.

Endemic: An organism that is common in the ecosystem at any time. Particularly applicable to pathogens.

Endotoxin: A poison produced by a microbial cell that is released following cell lysis.

Energy Budget: The balance between energy inputs and energy demands in an ecosystem.

Enteric: Dealing with human intestines.

Environment: The abiotic and biotic factors controlling the life of an organism.

Enzyme: An organic catalyst.

Epidemic: A large incidence of a disease in a population.

Epilimnion: The warm circulating upper level of a large body of water.

Eucaryote: A protistan cell having a distinct nucleus.

Euphotic zone: The zone in water where there is sufficient light for algal productivity.

Eutrophication: The enrichment of a body of water with nutrients.

Excretion: Waste products produced by a cell.

Exotoxin: A poison excreted by a microbial cell.

Exponential growth: The phase in the growth curve of bacteria in which the statistical population is growing at an exponential rate.

Exudate: Metabolic products excreted by microorganisms.

Facultative anaerobe: A microorganism capable of growing either aerobically or anaerobically.

Feedback inhibition: Inhibition of an enzyme caused by a product of that enzyme's activity on a substrate.

Fermentation: Anaerobic degradation of a substrate.

Fimbiae: Very small hairs on the surface of some bacterial cells.

Flagellum: A thread-like portion of many motile microbial cells that is responsible for the cell motility.

Flocculation: Aggregation of cells in liquid.

Food chain or food web: The sequence of organisms that feed on each other, starting at the phytoplankton and leading to very large organisms.

Gamete: A sexual cell that joins with another cell of opposite sex to form a zygote and ultimately a new organism.

Gene: A unit of genetic information.

Genus: A group of organisms containing closely related species.

Glucan: A carbohydrate polymer formed from glucose.

Gram stain: A technique used to differentiate bacteria. Gram-positive bacteria stain violet and Gram-negative bacteria stain red.

Growth curve: The curve representing the growth phases of bacteria during the life of a culture.

Habitat: The environment in which an organism lives.

Halophile: A microorganism that lives at high salinity.

Heterotroph: A microorganism that utilizes organic compounds for its energy and carbon requirements.

Homeostasis: The buffering capacity of an ecosystem that allows it to resist perturbations.

Hypolimnion: The cold, lower level of a large body of water.

Imhoff tank: An anaerobic sewage treatment tank in which the solids are withdrawn from the bottom of the tank.

Incubation: Growth of a microbial culture under specific environmental conditions.

Infection: Disease caused by growth of a microorganism.

Inhibition: Prevention of microbial growth.

Inoculation: Introduction of microorganisms into an environment or a culture medium.

Inoculum: The microorganisms inoculated into a medium.

Intracellular: Inside the cell.

Ion exchange: The use of chemicals, usually in columns, to remove an ion from water and replace it with another ion.

Lag phase: The phase of a microbial growth curve, directly after inoculation, when the cells are adapting to the new medium.

Lichen: A symbiotic association between an alga and a fungus.

Littoral: Living in the shallow waters of lakes or the sea.

Lysis: Cell destruction. This may be accomplished by autolytic or external enzymes or by artificial means.

Mannan: A carbohydrate polymer composed of mannose units.

Metabolism: The biochemical processes in a cell.

Metachromatic granules: Dark granules composed of polyphosphates found in many bacteria.

Micheles-Menten equation: An equation used to describe the kinetics of enzymatic activity.

Millipore filters: A series of membrane filters commonly used to remove bacteria. A typical bacterial filter has a pore size of $0.45\,\mu$.

Monotrichous: Having one flagellum.

Mutant: An organism with a changed characteristic resulting from a genetic change.

Mutualism: A benevolent association between two microorganisms that benefits both.

Mycelium: The network of hyphae that constitutes a mat of fungal growth.

Nekton: Swimming organisms, including fish.

Niche: The place of an organism is an ecosystem.

Nitrification: The biological oxidation of ammonia to nitrate.

Nitrogen fixation: The reduction of atmospheric nitrogen to cell nitrogen by microorganisms.

Nucleus: The portion of a cell that contains DNA.

Nutrient: A substrate used to stimulate growth.

Obligate: Essential for growth, as in *obligate anaerobe*.

Optimum: The range in pH, temperature, or other parameter at which growth or enzymatic activity is greatest.

Organic: A molecule that is of biological origin and contains carbon. The opposite is *inorganic*.

Oxidation: The chemical reaction that involves combination with oxygen or loss of electrons by an electron donor.

Parasitic: The ability to grow on a living host. May not necessarily be associated with disease.

Pasteurization: The treatment of milk, alcoholic beverages, or other materials by heat to destroy most of the microorganisms.

Pathogenic: Having the potential to cause disease.

Peritrichous: Many flagella surrounding a cell.

Pesticide: A chemical used to control pests. These may include insecticides, herbicides, and fungicides.

Petri dish: A glass or plastic dish with a cover used to grow microorganisms on solid media.

Pheromone: Externally produced chemicals that control behavior in other individuals of the same species. Allomones control behavior in individuals belonging to different species.

Photoautotroph: An organism capable of utilizing light energy for growth.

Photosynthesis: The series of biochemical reactions involving light energy, carbon dioxide, and water used by photosynthetic organisms for growth.

Phycosphere: The zone of water influenced by algal exudates.

Phytoplankton: The plant-like unicellular organisms in water. The animal-like unicellular organisms are *zooplankton*.

Polychaetes: Segmented worms. Mainly marine, living in muds, in both deep and shallow waters.

Polypeptide: A polymer formed either by synthesis from amino acid or by protein degradation.

Polysaccharide: A polymeric carbohydrate formed by synthesis from monosaccharides or disaccharides (sugars). Starch, cellulose, and glycogen are polysaccharides.

Predator: An organism that lives by eating other organisms, their prey.

Primary productivity: Production of photosynthetic organisms, the first step in the food chain.

Procaryotic: A protist with a simple structure. Specifically the nucleus has no nuclear membrane.

Proteins: Organic polymers composed of amino acids.

Proteolytic enzymes: Enzymes that degrade proteins.

Protoplasm: The material inside a living cell.

Pseudopodia: False feet formed by amoeboid protozoa.

Pure culture: A culture or medium containing one organism.

Radioactive isotope: A form of an element that emits radiation. Also called radionuclide.

Recalcitrant chemical: One that is not easily degraded by micro-organisms.

Reduction: A chemical reaction in which oxygen is removed or electrons are accepted by an electron acceptor.

Respiration: Aerobic metabolism in a cell. The energy for growth in respiratory processes is obtained by transfer of electrons from a substrate to oxygen as the ultimate electron acceptor.

Reverse osmosis: Desalination by filtration of water through a semipermeable membrane. Ultrafiltration is used to filter larger molecules.

Rhizobium: The bacterium associated with legumes in nitrogen fixation.

Rhizosphere: The zone of soil influenced by plant root exudates.

Saprophyte: A nonpathogenic microorganism.

Saturated hydrocarbon: A hydrocarbon in which there are no double or triple bond linkages. Therefore, the hydrocarbon is more stable. Contrast *unsaturated* hydrocarbon.

Secondary treatment: A process used to oxidize organic matter in wastes by conversion to bacteria under high aeration.

Self-purification: The ability of a body of water to rid itself of pollutants.

Septic tank: A tank used in rural areas for disposal of domestic wastes.

Septum (plural, Septa): The cross wall of a fungal hypha.

Sessile: An organism that is permanently attached to a surface.

Sludge: The microbial cells that settle out during aerobic sewage treatment.

Spartina: The dominant plant in a salt marsh community.

Species: A very closely related group of organisms within a genus.

Spermatophyte: Seed-forming plants.

Spore: A resistant form of a microorganism that can survive adverse conditions for very long periods of time.

Stability: The ability to resist perturbations.

Stationary phase: The phase in a growth curve in which the death rate equals the growth rate.

Sterile: Free of microorganisms.

Streeter-Phelps equation: A means of predicting the permissive organic load to a stream based on the deaeration and reaeration rates resulting in an oxygen sag curve.

Substrate: The molecule which reacts with an enzyme or which is used for growth.

Succession: The continuous replacement in time of one community by another. Ultimately leads to a climax stable community.

Symbiosis: Two organisms living closely together for their mutual benefit.

Thermocline: The sharp zone in a stratified body of water that separates the epilimnion from the hypolimnion.

Thermophile: An organism that has a high temperature optimum. Compare *mesophile* and *psychrophile*.

Toxin: A poison produced by an organism.

Trace element: A chemical element required in extremely small quantities by a microorganism.

Trophic level: The level of an organism in a community based on its feeding requirements. *Phytoplankton* are at the first level; *zooplankton* are at the second.

Vaccine: A culture of a pathogenic microorganisms in which the ability to cause disease has been eradicated by killing or attenuation of the cells. Inoculation of the treated cells provides immunity from the disease.

Virion: A viral cell.

Yeasts: A group of fungi that produce characteristic budding cells rather than hyphae.

Zygote: A stage in the life cycle of fungi produced by the union of two gametes.

INDEX